# Molecular Biology
# Biochemistry and Biophysics

# 11

P. Reeves

# The Bacteriocins

With 9 Figures

Springer-Verlag New York · Heidelberg · Berlin 1972

*Peter Reeves, Ph. D.*

Senior Lecturer in Microbiology
Department of Microbiology, University of Adelaide, Australia

ISBN 0-387-05735-8 Springer-Verlag New York Heidelberg Berlin
ISBN 3-540-05735-8 Springer-Verlag Berlin Heidelberg New York

# Preface

In recent years bacteriocins, especially colicins, have become widely known to molecular biologists as proteins with peculiar ways of killing bacteria. These same bacteriocins have been known for a long time to bacteriology for their unusual activity spectra and enormous variety. In this monograph I have attempted to bring together our detailed knowledge of those few bacteriocins which have already received attention from molecular biologists, and our less detailed but extensive knowledge of the variety of bacteriocins which exist.

The field has been reviewed in whole or in part, by several authors [FREDERICQ, 1957, 1964, 1965 (2); IVANOVICS, 1962; HAMON, 1965; REEVES, 1965 (2)]. These reviews have been very useful to the author, and readers will find further references in them, and sometimes alternative viewpoints.

We have already referred to bacteriocins as proteins, and in doing so have excluded many more complex antibacterial agents which resemble bacteriophages or their tails. In the author's view, these phage-like particles are probably not bacteriocins, but many authors include them within the definition; the more restrictive definition used here has meant omitting discussion of some excellent studies on what the present author would term defective bacteriophages.

In the first chapter we look at the discovery of bacteriocins and an outline of their classification. With this background we can discuss in Chapters 2 to 6 the chemistry, genetics and mode of action of the more intensively studied bacteriocins.

The properties of bacteriocins pose some general questions for biologists: what is their evolutionary origin and what role do they play in bacterial ecology? These questions are discussed, although regrettably not answered, in Chapter 7.

The Appendix is a catalogue of bacteriocins in all their bewildering variety. They are classified according to the producing strain, which is the best that can be done at present. Most of them have never appeared in the pages of the *Journal of Molecular Biology*, but is is hoped that this rather detailed compilation will help the reader to place those which have in a more general context. The variety is in itself of interest to molecular biology, but at present we can only speculate as to the mode of action or biological significance of most of them.

Even in writing on such a limited topic, I have found it necessary to ask colleagues and friends for help, and this has been generously given. It gives me pleasure to acknowledge the help I have received from those who sent me manuscripts and have discussed their work with me. I particularly wish to express my gratitude to Dr. G. F. SPRINGER, who has been very patient as editor, and to Drs. Y. HAMON, D. HARDY, V. KRISHNAPILLAE, S. E. LURIA, R. NAGEL DE ZWAIG, M. NOMURA, and B. L. REYNOLDS, who all read the manuscript in whole or in part and whose criticisms were invaluable.

P. REEVES

# Contents

## Chapter 4  The Properties of C-Factors

## Chapter 5  Mode of Action — The Adsorption of Bacteriocins

Chapter 6 **Mode of Action — The Biochemical Lesion**

# Chapter 1

# The Colicins and Other Bacteriocins

## The Discovery and Classification of Colicins

In 1925 ANDRE GRATIA published his observations and studies on an antibacterial substance which we know as a colicin. He showed that one strain of *Escherichia coli* (called V because it was a virulent strain) produced a substance which was bactericidal for *E. coli* ø (also known as strain CA81). Gratia's strain V was later renamed CA7 but the original name survives in colicin V, the name now given to the substance it produces.

This colicin V could be demonstrated in several ways. If a broth culture of *E. coli* V was sterilised by filtration, then the filtrate, even if diluted a thousandfold in broth, would not support growth of *E. coli* ø (GRATIA, 1925, 1932). In later work the double-layer technique has been widely used to demonstrate colicin activity (FREDERICQ, 1948). The colicin-producing strain is stabbed into a nutrient agar plate and incubated 24 or 48 h. The bacterial growth is then killed by exposure to chloroform vapour and a second layer of agar, incorporating a few drops of a culture of sensitive organisms, is poured over the first. After further incubation the sensitive organism grows as a confluent lawn of growth, interrupted only where it overlaps growth of a colicin-producing strain. The method of application of the two strains can, of course, be varied and may for instance consist of streaks at right angles to each other, whereby several potentially sensitive strains can be tested on one producing strain.

GRATIA's publications (1925, 1932) on colicin V-CA7 included observations on its stability to heat and chloroform and its dialysability, and further showed that it could be precipitated by acetone. However, we must now move to the work of the late 1940's done by FREDERICQ, initially with GRATIA. This second phase of the work on colicins rapidly extended the picture to include many strains of *E. coli* and other genera of Enterobacteriaceae. GRATIA and FREDERICQ (1946) found that about 20% of *E. coli* strains could produce substances active on their sensitive strain *E. coli* ø and they also found that about 50% were sensitive to colicin V-CA7. When FREDERICQ [1946 (1)] compared the activity spectra of 7 colicinogenic strains he found that each inhibited from 1 to 5 of 10 sensitive strains and that no two activity spectra were identical.

It was obvious that many different colicins existed and these could be classified in part at least by their activity spectra, although we should take note of Fredericq's caution that some strains may produce more than one colicin, the activity spectra of such strains being the sum of those of two or more colicins. FREDERICQ then undertook an extensive study of colicinogenic bacteria and in 1948 in his review considers

Table 1.1. *The cross-resistance of colicin resistant mutants.* [*After table 9 of* FREDERICQ (1948)]

| Colicin used for selection | Total | Number of mutants resistant to colicins | | | | | | | | | | | | | | | | |
|---|---|---|---|---|---|---|---|---|---|---|---|---|---|---|---|---|---|---|
| | | A | V | B | C | D | E | F | G | H | I | J | K | S1 | S2 | S3 | S4 | S5 |
| V | 29 | 29 | — | — | 3 | — | 5 | 4 | 1 | — | — | — | — | — | — | — | — | — |
| A | 13 | — | 13 | — | — | — | 1 | — | 4 | — | 3 | 5 | 3 | 5 | 3 | 1 | — | 3 |
| B | 65 | 2 | 1 | 65 | 30 | 17 | — | — | — | 3 | 3 | 1 | 3 | — | 1 | 1 | 2 | 1 |
| C | 22 | — | — | 8 | 22 | 11 | — | — | — | — | — | — | — | — | — | — | — | — |
| D | 23 | — | — | — | 8 | 23 | — | — | — | — | — | — | 1 | — | — | — | — | — |
| E | 5 | — | 3 | — | — | — | 5 | 5 | — | — | 1 | 4 | — | — | 4 | 2 | — | 4 |
| F | 6 | — | 6 | — | — | — | 6 | 6 | — | — | — | 6 | — | — | 4 | 5 | — | 4 |
| G | 8 | — | 1 | — | — | — | 3 | 2 | 8 | 2 | 2 | 1 | — | 1 | 1 | 1 | — | 1 |
| H | 1 | — | — | — | 1 | — | — | — | — | 1 | — | — | 1 | — | — | — | 1 | — |
| I | 11 | — | 1 | — | — | — | 9 | 9 | 1 | — | 11 | 5 | — | — | 1 | 1 | — | 1 |
| J | 13 | — | 9 | — | 1 | — | 1 | — | — | — | 1 | 13 | — | — | 9 | 8 | 1 | 10 |
| K | 20 | — | 3 | — | 1 | — | 2 | — | 5 | — | 8 | 2 | 20 | 4 | 2 | 2 | 2 | 5 |
| S1 | 13 | — | 4 | — | — | — | 3 | 3 | 2 | — | 2 | 3 | 2 | 13 | 2 | 2 | 3 | 4 |
| S2 | 12 | — | 2 | — | — | — | 3 | 3 | — | — | — | 2 | 1 | 2 | 12 | 6 | — | 4 |
| S3 | 7 | — | 3 | — | — | — | 3 | 3 | — | — | — | 2 | 1 | — | 6 | 7 | — | 2 |
| S5 | 4 | — | 2 | — | — | — | 1 | 2 | — | — | 2 | 4 | 1 | — | 4 | 1 | — | 4 |

252 resistant mutants selected for resistance to the colicin indicated in the left-hand column were tested for resistance to each of the 17 colicins recognized at that time. The mutants were derived from 12 different sensitive strains

in some detail the action of 88 colicinogenic strains using 316 strains as indicators. He was able to recognize 17 distinct patterns of activity spectra, after allowing for some which were presumed to be due to the production of two colicins, and concluded that there were at least 17 distinct colicin types. (The 88 strains were chosen from 254 colicinogenic strains, selecting as much apparent diversity as possible.)

An alternative way to study the variety of colicins is to catalogue their activity spectra against colicin-resistant mutants derived from a sensitive strain. GRATIA (1932) showed that it was easy to isolate mutants from the inhibition zones produced by CA7 on *E. coli* ø and FREDERICQ [1946 (3)] derived 8 mutants of *E. coli* ø each resistant to a different colicinogenic strain. The 8 mutants were tested for their sensitivity to 3 colicinogenic strains and fell into 4 groups. Thus the mutations showed considerable specificity and can be used to distinguish different colicins and to classify them.

Two factors make the classification of colicins by resistant mutants less simple in practice than it might seem in principle. Firstly some strains produce several colicins and a mutant selected as resistant to such a strain will be resistant to two different colicins. It is sometimes possible to resolve such a situation by picking a mutant from the edge of the inhibition zone, which, while resistant to the more diffusible colicin, may remain sensitive to a less diffusible colicin which does not reach the edge of the zone. Such a mutant will remain sensitive to the strain used to isolate it but show a smaller zone of inhibition [FREDERICQ, 1946 (5); SMARDA, 1960].

A second difficulty arises because some mutations confer resistance to two or more colicins which can nonetheless be readily distinguished by other mutants. Table 1.1, which is taken from Fredericq's review (1948), shows that mutants which do not show complete specificity are quite common. Nonetheless FREDERICQ was in general able to obtain mutants which were specifically resistant to colicins of one of the 17 types recognized earlier on the basis of activity spectra. There was however one major exception and FREDERICQ later [1950 (1)] amalgamated colicin types, E, F, J, S2, S3 and S5 to form colicin type E, because of the very frequent occurrence of mutants conferring cross-resistance to colicins of these 6 types.

## Specific Receptors in Colicin Classification

FREDERICQ [1946 (3), 1948] suggested that colicin resistance involves the loss of a surface receptor specific for colicins of that type. Later work, which we shall also discuss in Chap. 5, has confirmed that some colicins at least do adsorb to specific receptors which can be lost by mutation.

Mutants selected by colicins of type E have been used in several studies (CLOWES, 1965; HILL and HOLLAND, 1967; NAGEL DE ZWAIG and LURIA, 1967; NOMURA, 1964; NOMURA and WITTEN, 1967; REEVES, 1966) and two classes of mutants are distinguished. Those of one class confer resistance to all colicins of type E and also phage BF23, thought by FREDERICQ [1951 (3)] to adsorb to the same receptor. They do not adsorb colicin and map at the cer locus on the genetic map (PFAFF and WHITNEY, 1971). Mutants of the other more heterogeneous group may not be resistant to all type E colicins, may resist some colicins of other types, are usually sensitive to phage BF23, and still adsorb colicin. REEVES (1966) extended studies with these two types of mutants to many colicins of type E and found that mutants of the former class showed absolute specificity for colicins of this type, resisting all of them as well as BF23. In this instance then we have a well-defined class of "receptor" mutants and might reasonably conclude that colicin E is in fact defined by a specific receptor. However, as we shall see in Chapt. 5, even here there is still some doubt. Similarly colicin K-resistant mutants have been found to lack receptor (NAGEL DE ZWAIG and LURIA, 1967). Unfortunately, there have been almost no studies on the nature of the cross-resistance observed for other colicins although there are considerable genetic data available concerning cross-resistance between colicins of types I, V and B and phage T1 (GRATIA, 1964).

The classification set up by FREDERICQ, based on resistant mutants, is now in general use. It seems likely that this classification is indeed based on receptor specificity but we should remember that this is not established experimentally in most cases.

In addition to these two methods for classifying the large number of colicinogenic strains discovered, FREDERICQ also used the diameter and appearance of the inhibition zone produced on a standard indicator strain and such properties as dialysability and heat stability. More recently LIKHODED [1963 (2)] has made a comparative study of such properties using colicins of types B, D, E, G and V. In some cases he found a good correlation with the classification into types defined by resistant mutants.

# Terminology

Several authors have recently pointed out the necessity to carefully define a colicin used in any particular study, and until much more is known about the nature of colicins, the only completely satisfactory description is the name of the producing strain. Just as each bacteriophage strain is given its own trivial name, so each colicin needs a trivial name. Thus the colicin of type V produced by *E. coli* CA7 is known as colicin V-CA7.

Most of the colicinogenic strains in general use have been given to the scientific community by Prof. FREDERICQ. His strains are in several series, prefixed CA, K, S, P, GEI or N. Because the several series of strains isolated and given *different* alphabetical prefixes all number from 1, the prefix is necessary for a full identification.

# Subcategories of Colicin Types

Some of the colicin types defined by resistant mutants have been shown to be heterogeneous when studied in other ways, and we noted before the variation in activity spectra.

Colicins of type E are frequently determined by genetic entities known as C-factors, which are readily transferred to other strains of *E. coli*, rendering them colicinogenic. FREDERICQ [1956 (2)] found such newly colicinogenic strains to be immune to the colicin they now produce and some but not all other colicins of type E. He obtained several colicinogenic derivatives from *E. coli* K12 by allowing transfer of a C-factor from different colicinogenic strains. Using two such derivatives he was able to distinguish 3 different groups of colicins among those of type E, each group having a specific activity pattern on the two strains *E. coli* K12 (E1-K30) and K12 (E2-K317), these being *E. coli* K12 carrying the C-factors for colicin E from *E. coli* K30 and *E. coli* K317 respectively. Colicins, like E1-K30, unable to act on *E. coli* K12 (E1-K30) are known as type E1 while another group are unable to act on K12 (E2-K317) and are called E2, the colicins of K30 and K317 being the type colicins of these two subtypes. Other colicins acted on both strains used and were called type E3. Recently the C-factor E3-CA38 has been transferred to *E. coli* K12 [HERSCHMAN and HELINSKI, 1967 (2)] but the immunity of this strain has not been reported. HAMON (1964) described a fourth type, E4, comprising a colicin produced by a strain called *E. coli* H. *E. coli* K12 (E4-H) is immune to colicins of type E1 and E2 (E3 is not mentioned) and conversely is not active on bacteria carrying E1 or E2 (the strains used were not stated).

The subclassification of colicins into E1, E2 and E3 has received considerable prominence because they have been shown to have very different modes of action (see Chap. 6). However, we have to be cautious in our use of this classification as certain anomalies have been discovered, the C-factor E2-P9, for instance, conferring on *E. coli* K12 an immunity distinct in some respects from that conferred by the C-factor E2-K317 (LEWIS and STOCKER, 1965).

The colicins of type E have also been subdivided on the basis of cross-resistance conferred by mutants which do not involve the receptor (REEVES, 1966). Such mutants may confer tolerance (so called to distinguish it from the resistance due to receptor loss) to only some of the colicins of type E and once again the colicins may be classified on this basis, but the molecular basis for this subdivision is not known. It should be noted that the two methods for subdividing the colicins of type E are in partial agreement but do not give identical results. The data on which these two classifications of type E are based will be discussed more fully in Chap. 6. Colicins of type E also differ in their activity spectra; this is discussed in the Appendix.

Colicins of type I have been subdivided into Ia and Ib on the basis of immunity conferred by certain C-factors (STOCKER, 1966). C-factors of type Ib not only confer immunity to colicins of type Ib, but also cause infections by phages BF23 and T5 to abort without release of functional phage (STOCKER, 1966; STROBEL and NOMURA, 1966; TAIZO and OZEKI, 1968). In this case the difference can also be detected by electrophoresis, three colicins of type Ia behaving differently to three of type Ib (SMITH, 1966).

The colicins of type V have also been subdivided but on a different basis. HAMON (1964) observed that some strains were sensitive to only some colicins of type V, and he was able to subdivide these colicins on the basis of their activity spectra and also using the cross-immunity pattern. This subclassification has not yet come into general use, perhaps because the strains used are not themselves in general use, much less work being done on colicins of type V than of type E. MACFARREN and CLOWES (1967) introduced an independent subdivision of colicin V C-factors.

The three examples of subdivision of colicin types show that there is considerably more heterogeneity among the colicins than is indicated by the receptor classification, but as yet the fundamental basis for these subdivisions is not known. It seems certain that much more heterogeneity is yet to be discovered in the colicins and this may well be of value in elucidating their mode of action.

## Colicins as Bacteriocins and Other Groups of Bacteriocins

Colicins differ from most of the previously known antibiotics in certain respects, their principal characteristic being the restriction of activity to strains of species related to the producing species and particularly to strains of the same species. The classical antibiotics usually have a much wider activity spectrum, and even when the activity is restricted, it does not show this preferential effect on closely related strains. Fredericq's original studies on colicins involved only strains from the family Enterobacteriaceae, although within this family he found that the activity of colicins (mostly produced by *E. coli* and *Shigella*) was less frequent against genera such as *Proteus* than against *E. coli* and *Shigella*. The only type colicin which has been shown to have any activity outside this family is colicin G, which acts on strains of *Pasteurella* (SMITH and BURROWS, 1962). In addition to these data there are several surveys such as that of HALBERT [1948 (1, 2, 3)], in which a few strains of unrelated species were used that were never sensitive to colicins, and there are several references in Hamon's work on *other* groups of bacteriocins to the lack of activity of colicins on the indicator strains for the newly described groups of bacteriocins. However, the activity of colicins may be more extensive, as BLACKFORD, PARR and ROBBINS (1951) and COOK, BLACKFORD, ROBBINS and PARR (1953) found some *E. coli* and other coliforms to be active against *Vibrio metschnikovii*, *Pseudomonas*, *Rhodospirillum*, and *Neisseria*, and even against several genera of gram-positive bacteria. Unfortunately the colicins produced by these strains were not typed, and so we do not know if there is any correlation between colicin type and this extended activity. FREDERICQ [1958 (2), 1963 (1)] considers this activity outside the Enterobacteriaceae to be due to factors other than colicins.

More recently PAPAVASSILIOU has identified two new colicin types, X and L [HAUDUROY and PAPAVASSILIOU, 1962 (1); PAPAVASSILIOU, 1961 (1)], the first being active against 77% of *E. coli* and also some other Enterobacteriaceae but none of 77 strains of other bacterial families tested. Likewise colicin L is active against *E. coli* (22%) but in this case no other genera within the family proved sensitive and none of 95 strains from outside the family Enterobacteriaceae were sensitive. For these two colicins at least the activity seems to be confined to strains closely related to the producer. Thus despite the lack of any systematic studies on the possibility of colicin action outside the Enterobacteriaceae, it seems reasonable to say that such activity must be rare even if present at all.

The second notable characteristic of colicins is of course their occurrence in the variety of different types which we have discussed above, each active on a particular range of strains within the family Enterobacteriaceae. Other characteristics of colicins, such as their chemical nature and the nature of their genetic determination, are also unusual and we shall discuss these in detail later, but such characteristics are not suitable for the initial identification of a colicin because of the greater amount of work involved in their study.

The word bacteriocin was coined by JACOB, LWOFF, SIMINOVITCH and WOLLMAN (1953) to include colicins and related substances produced by other groups of bacteria. On the basis of an activity restricted to closely related bacteria and the further criterion of the existence of a variety of different types, several distinct families of bacteriocins can now be recognized and these are discussed in some detail in the appendix. Many of them have been described since 1959 by HAMON and his co-workers as a result of an extensive search. Several of the bacteriocin families from the gram-negative bacteria are closely analogous to colicins in both properties. In addition to the bacteriocins of other genera of Enterobacteriaceae, we have in particular the pyocins and fluocins (from *Pseudomonas pyocyanea* and *P. fluorescens* respectively) which fit closely to the pattern known from the colicins (see Appendix).

However, even within the gram-negative bacteria we have some families (the pesticins, named after *Pasteurella pestis*, for example) which do not fit this prototype pattern so closely and the bacteriocins of the gram-positive bacteria often deviate markedly from the colicin model, having wide spectra and only a very limited number of different types (see Appendix).

Although the study of activity spectra serves to identify probable bacteriocins, we will be able to discuss the relationship of these "atypical" bacteriocins more meaningfully after we have looked at the chemical and other experimental studies which have been carried out with bacteriocins. It is for this reason that we shall defer further discussion of the diversity of bacteria which produce bacteriocins until the end of the book and look next at what is known of the relatively few which have been studied in detail.

In writing a book on bacteriocins, it has been necessary to use some guiding principle as to what constitutes a bacteriocin. The only real difficulty arises in distinguishing colicins from bacteriophages, and in particular from defective bacteriophages. We shall see later in Chap. 2 that some of the "bacteriocins" which have been described have a structure which clearly shows that they are defective bacteriophages. In other cases the evidence is only indicative. Since a full discussion of bacteriophages is beyond the scope of this book, a decision had to be made on where and how to draw the line between bacteriocins and bacteriophages. In general any agent which is now known to have a structural resemblance to a bacteriophage has not been discussed in this book, except in passing. However, many of the agents discussed are of unknown structure, and it is likely that some will eventually be shown to be defective bacteriophages. It is the author's contention that bacteriocins and defective bacteriophages are quite distinct, if not always distinguished in practice. However, it should be noted that this is a far from universally accepted view, and other authors would include as bacteriocins agents considered here as defective bacteriophages.

# Chapter 2

# The Chemistry

## Chemical Studies on Colicins and Related Bacteriocins

### The First Observations

Bacteriocins are very potent antibacterial agents, and it is not surprising that the early work included some chemical studies. These showed that several colicins had in general one or more of the following properties: 1. inactivation by proteolytic enzymes; 2. retention by cellophane dialysis membranes; and 3. precipitation by agents such as acetone or ammonium sulphate, known to precipitate proteins. Most of the authors concluded, with varying degrees of conviction, that the colicins involved were peptides or proteins (DEPOUX and CHABBERT, 1953; GARDNER, 1950; GRATIA, 1932; HALBERT and MAGNUSON, 1948; HEATLEY and FLOREY, 1946; JACOB, SIMINOVITCH and WOLLMANN, 1952). Some of the colicins used have never been classified in Fredericq's scheme but others included examples of types D, E and V. However, it was not until the work of GOEBEL and his colleagues on colicin K-K235 that a colicin was purified and adequately characterised.

### Colicin K-K235

The study of colicin K-K235 began in 1958 when GOEBEL and BARRY prepared an active material from culture supernatants. After growth in a fully dialysable medium they removed the bacteria by centrifugation and collected the macromolecular products in the supernatant by dialysis. They then separated the active material by various fractionation procedures including precipitations by ethanol, a chloroform-octanol mixture and ammonium sulphate.

This first preparation of the colicin was essentially homogeneous on free electrophoresis or on zone electrophoresis at various pH values, and consisted of a complex possessing the serological specificity of the O antigen, and containing protein, carbohydrate and lipid. This complex clearly resembles the "Boivin antigen" (BOIVIN and MESROBEANU, 1933, 1935; KABAT and MAYER, 1961) and contains perhaps 30% protein. Even in this first paper it was shown that the colicin activity resided in only part of the complex, since by two different procedures they were able to isolate from it all the colicin activity in a fraction containing about 85% protein and representing only 10% of the dry weight of original complex.

MIYAMI, ICHIKAWA and AMANO (1959) showed that under some circumstances the colicin activity was released from cells, unassociated with O antigen, but it was not until much later that colicin K-K235 was obtained as a pure protein free of any carbohydrate or lipid (DANDEU and BARBU, 1967; JESAITIS, 1967, 1970; KUNUGITA and MATSUHASHI, 1970; TSAO and GOEBEL, 1969). DANDEU and BARBU (1967), TSAO and GOEBEL (1969) and KUNUGITA and MATSUHASHI (1970) obtained the colicin as a protein after inducing *E. coli* K235 with mitomycin C induction and JESAITIS (1967, 1970) also used mitomycin C induction, but this time of a *Proteus mirabilis* strain which was carrying the C-factor derived originally from *E. coli* K235. JESAITIS (1970) purified the colicin from a culture supernatant after precipitation at pH 4.3. The purification involved precipitation between 33 and 66% saturation of ammonium sulphate followed by sephadex, DEAE-sephadex and CM-sephadex chromatography. The resulting purified colicin contained less than 0.3% phosphorus, lipid or carbohydrate and 90% of it could be accounted for as amino acids. It appeared to be homogenous on ultra-centrifugation ($S_{20, w}$ was 2.81 and 2.86 under two different conditions). The intrinsic viscosity indicated that the molecule was asymmetric and a molecular weight of 44,900 was calculated using the sedimentation data, partial specific volume and its behaviour on Sephadex G 200. Although the material gave a single boundary on free electrophoresis, two major and two minor components were recognised by isoelectric focusing and three components could be identified by acrylamide gel electrophoresis. Two major components separated by isoelectric focusing both have colicin activity.

We referred earlier to the association of colicin K-K235 with a complex which included the carbohydrate O antigens. These very good antigens have been studied in great detail (see LUDERITZ, STAUB and WESTPHAL, 1966 for review), but the proteins which are often associated with them have generally received little attention.

GOEBEL and BARRY (1958) obtained their protein-rich fractions from the complex by either phenol-water extraction or by precipitation with a complex mixture which they called "lower phase solvent". Neither procedure would have disrupted covalent bonds and the association may well have been due to hydrophobic bonds. However, these protein-rich fractions still contained some carbohydrate and lipid (about 3% and 6% respectively) and the nature of this association is not known at all. It has not been possible to separate an active pure protein from the complex by DEAE chromatography (GOEBEL, 1962) or indeed by any other means.

The K-K235 C-factor can be transferred to *Shigella* and *Proteus* as well as to other strains of *E. coli* and the colicin produced by *E. coli* K12 (K-K235) or *Shigella sonnei* (K-K235) is present in a complex containing the O antigen serological specificity and a high proportion of carbohydrate and lipid as well as a protein (AMANO, GOEBEL and MILLER-SMIDTH, 1958; HINSDILL and GOEBEL, 1964, 1966). The active materials prepared either from *E. coli* K12 (K-K235) or from *S. sonnei* (K-K235) were virtually indistinguishable in their chemical analysis from the corresponding material derived from the respective noncolicinogenic strains. Similarly the complexes from parent and daughter strains were indistinguishable serologically if assayed by quantitative precipitation or gel precipitation, although of course the materials derived from the *E. coli* K12 strains were readily distinguished from those derived from the *S. sonnei* strains. Likewise a material derived from a noncolicinogenic derivative of *E. coli* K235 was indistinguishable on the same criteria from the active material derived from the parent colicinogenic strain (RUDE and GOEBEL, 1962).

Thus the presence of colicin does not significantly alter the overall composition of the O antigen complex, and the colicin probably constitutes only a small proportion of the total protein. This is supported by the fact that the pure colicin from

*Proteus* is about a hundredfold more active than the protein-rich fractions from *E. coli* K235 (GOEBEL and BARRY, 1958; JESAITIS, 1970).

Antisera to the O antigen-colicin complexes of *E. coli* K235 or the colicinogenic derivative of *S. sonnei* contained colicin-neutralizing antibodies which remain after adsorption of the precipitating antibody with the noncolicinogenic strain (AMANO *et al.*, 1958; HINSDILL and GOEBEL, 1964, 1966). These antibodies will neutralise the colicin whatever its state, but will precipitate the colicin only when it is free and not when it is complexed with O antigen. These neutralising antibodies were further shown by TSAO and GOEBEL (1969) to be heterogeneous and variable in their ability to precipitate the free colicin.

## Colicins V-K357, A-CA31 and SG710

No other colicin has been studied in the same detail as K-K235, but 3 others, A-CA31, V-K357 and a colicin of unknown type, SG710, have been shown to be, at the very least, closely associated with the O antigen of the producing strain.

The composition of V-K357 (HUTTON and GOEBEL, 1961, 1962) shown in Table 2.1 is within the range for O antigens, gives the serological tests expected of an O antigen and is toxic for mice. However, the purification in this case was much less extensive than was used for K-K235, consisting of ethanol precipitation of the active component from the culture supernatant, and passage through DEAE cellulose to remove coloured material. The product nonetheless was electrophoretically and serologically homogeneous.

Colicin A-CA31 was also purified from the supernatant of a culture grown in a completely dialysable medium (BARRY, EVERHART, ABBOT and GRAHAM, 1965; BARRY, EVERHART and GRAHAM, 1963). In this case, after clarifying the culture supernatant by centrifugation, the colicin was purified by dialysis followed by precipitation three times at 4 °C with one volume of ethanol. The purified material gave a single broad peak on ultracentrifugation indicating that it was heterodisperse. It had an $S_{20}$ value of 4.6, indicating a molecular weight of $5 \times 10^4$, if certain assumptions were made. The chemical composition (Table 2.1) was again that expected for an O antigen. BARRY *et al.* (1965) also studied the colicins of type A produced by two more strains, and in each case colicin activity was found associated with material resembling an O antigen, in composition and serologically. The three organisms were serologically quite distinct when measured by agglutination, yet antisera against any one could neutralize the colicin activity of the others. The authors conclude that the colicin activity must reside in a molecule quite distinct from the O antigen although probably closely bound to it. DANDEU and BARBU (1968) also purified a colicin of type A.

The colicin from *E. coli* SG710 was studied and purified by NUSKE, HOSEL, VENNER and ZINNER (1957) but, unfortunately, never classified in Fredericq's scheme. The purified colicin again had the composition of an O antigen and the molecular weight was estimated to be between $10^6$ and $10^7$.

## Colicins of Type E

Colicins E1-K30, E2-CA42, E2-P9 and E3-CA38 have been purified and studied to various extents. Their production is induced by either ultraviolet irradiation or

Table 2.1. *Chemical composition of various colicins and megacin*

| | % Nitrogen | % Phosphorus | % Carbohydrate | % Lipid | % Protein | Reference |
|---|---|---|---|---|---|---|
| Colicin K (*E. coli* K235) | 5.0 | 1.4 | 26 | 10 | 30 | 2, 4 |
| Active protein-rich fraction of above | 14.0 | 0.4 | 6 | 3 | — | 2 |
| Colicin K *S. sonnei* (K-K235) | 11.1 | 1.0 | 4.3 | 7.6 | 70 | 4 |
| Somatic antigen- (*S. sonnei* col⁻) | 10.6 | 1.1 | 6.7 | 9.8 | 68.4 | 4 |
| Colicin K (*P. mirabilis* (K-K235) | 15.8 | <0.3 | <0.3 | <0.3 | 89 | 7 |
| Colicin E 2 *E. coli* (E 2-P9) | | <1.0 | <1.0 | | 98 | 3 |
| Colicin E 3 *E. coli* (E 3-CA38) | | <1.0 | <1.0 | | 99 | 3 |
| Colicin E 1 *E. coli* (E 1-K30) | | 0 | <1.0 | | 97 | 8 |
| Colicin SG710 | — | | 22 | 18 | 50 | 9 |
| Colicin V-K357 | 3.8 | 2.02 | 43 | 11.3 | | 6 |
| Colicin A-CA31 | 8.35 | 2.65 | 8.82 | 14.3 | 40 | 1 |
| *C. freundii* A-3653 | 4.09 | 5.40 | 25.25 | 15 | 20 | 1 |
| *E. coli* A-1077 | 8.61 | 4.45 | 17.18 | 12 | 42 | 1 |
| Megacin A-216 | 15.4 | <0.5 | <0.5 | — | — | 5 |

*References:* 1. Barry et al., 1965; 2. Goebel and Barry, 1958; 3. Herschman and Helinski, 1967 (2); 4. Hinsdill and Goebel, 1964, 1966; 5. Holland, 1961; 6. Hutton and Goebel, 1961, 1962; 7. J. Jesaitis, 1970; 8. Schwartz and Helinski, 1968; 9. Nuske et al., 1957.

mitomycin. The amount of colicin released into the medium is variable and the colicin has either been purified from the medium (REEVES, 1963 for E2-CA42) or after extraction from the collected cells by 1.0 M NaCl [HERSCHMAN and HELINSKI, 1967 (2) for E2-P9 and E3-CA38]. In both cases further purification was by ammonium sulphate precipitation, DEAE chromatography, and either CM cellulose or CM sephadex chromatography.

Colicins E2-P9 and E3-CA38 were obtained as proteins, homogeneous by poly-acrylamide gel electrophoresis and analytical ultracentrifugation [HERSCHMAN and HELINSKI, 1967 (2)]. They were both simple proteins of about 60,000 M.W. The amino acid compositions of the two colicins were similar but not identical. The iso-electric point of colicin E3-CA38 was pH 6.64 while colicin E2-P9 showed two components with isoelectric points of pH 7.63 and 7.41 by isoelectric focusing. Immunochemical analysis showed the two colicins to share some but not all of their antigenic determinants. The purest preparations of colicin E2-CA42 comprised only about 70% protein and contained some carbohydrate (REEVES, 1963). However, the molecular weight of about 60,000, calculated from the estimates of sedimentation coefficient and diffusion coefficient obtained from analytical ultracentrifugation, suggests that all three colicins may be similar.

Colicin E2-P9 has been shown to be labile at low concentrations in saline in the absence of some other protein such as bovine serum albumen (MITUSI and MIZUNO, 1969). This is presumably due to denaturation, and suitable precautions must be taken when assaying pure colicin preparations.

Colicin E1-K30 has been purified by SCHWARTZ and HELINSKI (1968) and found to be a protein of 55,000 M.W. Its amino acid composition differed considerably from that of E2-P9 and E3-CA38.

## Colicins of Type I

Colicins Ib-P9 and Ia-CL223 have both been partially purified after extraction from sonicated cells, by ammonium sulphate precipitation and DEAE cellulose chromatography (LEVISOHN, KONISKY and NOMURA, 1968). Molecular weights of 50,000 were estimated for both colicins from data on sedimentation of the bio-logical activity on sucrose gradients. Both colicins were inactivated by trypsin.

An unspecified colicin of type I was partially purified by KEENE (1966) and some of its properties studied. The material comprised about 60% protein, 6% carbo-hydrate, 5.5% lipid and 0.65% phosphorus. However, there was no indication that the material was very pure.

## Cloacin DF13

The cloacin DF13 has been purified from the supernatant of an induced culture by ammonium sulphate precipitation and ion exchange sephadex chromatography (DE GRAAF, GOEDVOLK-DE GROOT and STOUTHAMER, 1970; DE GRAAF, TIEZE, BONGA and STOUTHAMER, 1968). It was a simple protein of about 56,000 M.W.

## Morganocin MB336

This bacteriocin from *Proteus morganii* has been purified from the supernatant of an induced culture by ammonium sulphate precipitation and column chromatography

(SMIT, DE KLERK and COETZEE, 1968). It comprises 88% protein and 8% carbohydrate with no lipid or phosphorus detected. It has a sedimentation coefficient of 4.O S and is heat labile.

## Other Colicins

A bacteriocin produced by a strain of *Salmonella bergedorf* has been studied recently by ATKINSON (1967) and although she refers to it as a Salmonellin because it is active only on *Salmonella* and not on *E. coli* strains, we will provisionally refer to it as a colicin (see Appendix). This colicin is of particular interest in that it is very readily dialysable through sheets of cellophane or Visking dialysis tubing placed on the surface of agar plates, both when the bacteriocin is produced by *Salmonella* placed on top of the membrane and when this colicin is first extracted from agar plates on which the producing organism has grown and then placed on top of the membrane. It appears, both from its zone size and the extent to which it diffuses through dialysis membrane, to be of much lower molecular weight than any other known colicin and further chemical analysis will be very interesting.

## The Colicins Compared

It is difficult at this stage to say how different are those colicins found associated with the O antigen from those which are not. We know that colicin K-K235 is released free of any association with O antigen, after mitomycin induction (TSAO and GOEBEL, 1969). The colicin K-K235 produced by *P. mirabilis* (K-K235) and shown to be a pure protein was also induced by mitomycin C (JESAITIS, 1970), and JESAITIS has pointed out that the colicins of types E1, E2 and E3 studied by Helinski's group and shown to be pure proteins, were also induced by mitomycin C. With the exception of *E. coli* K235, the colicin synthesised by the same strain in the absence of induction has not been studied. JESAITIS (1970) suggests that on induction too much colicin is synthesised to be able to associate with the O antigen complex and the excess remains as free protein. However, it should be noted that even after induction colicins of type E may not be released into the medium but can be extracted as proteins from the cell surface without lysis, suggesting a rather labile association with cell wall components. It is not known if the O antigen complex derived from colicin E producers possesses colicin activity.

The conflicting data in the literature on the dialysability of colicin V may be due to its being released either free, or in association with O antigen as a high molecular weight complex. GRATIA (1925) and FREDERICQ [1963 (1)] found colicin V to diffuse from a colony of a producing strain through cellophane into agar, and HEATLEY and FLOREY (1946) also found their partially purified preparation of a colicin of this type to be freely dialysable. On the other hand HUTTON and GOEBEL (1961, 1962) found V-K357 to be nondialysable and used this property in purification. Some of the differences may be due to the use of different producing strains and purification procedures but this would seem not to account for all of it and CLOWES (personal communication) found that particular colicins of type V which were dialysable if measured by Gratia's technique were not if culture supernatants of the same strain were tested. BARRY *et al.* (1965) report similar findings for D-CA23 suggesting that it too can exist in two forms.

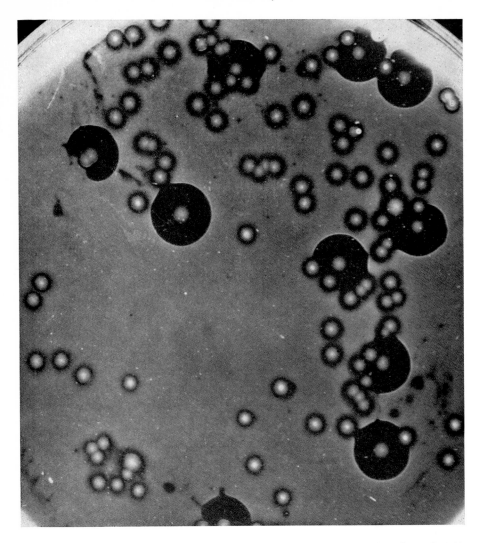

Fig. 2.1. *The effect of receptor on the inhibition zone of a colicinogenic strain.* Strains of *E. coli* K 12, either sensitive or resistant to colicin I, were made (I-P9)$^{+}$. A mixture of the two was plated on nutrient agar, incubated, the colonies killed by chloroform vapour and the plate overlaid with soft agar containing an indicator bacterium. The colicinogenic bacteria which lack receptor (from the resistant strain) give much larger inhibition zones. [From MONK and CLOWES, 1964 (1)]

In the case of colicin V we have the further discrepancy that while V-K357 prepared by HUTTON and GOEBEL is toxic for mice (LD50, 0.9 mg) as would be expected of an O antigen, V-CF1 prepared by HEATLEY and FLOREY was not only dialysable but not toxic for mice, which survived intravenous injection of 20 mg. This lack of toxicity was confirmed by BRAUDE and SIEMIENSKI (1964, 1965), who also used V-CF1. Again the differences may be due to strain differences or due to differences in the purity of the colicins used, but the possibility also remains that a given colicin activity is only sometimes associated with O antigen. The particular form obtained from a culture containing both would depend in part on the purification procedures used.

MONK and CLOWES [1964 (1)] found that the size of inhibition zone produced by *E. coli* K12 (I-P9) depended on the presence or absence on the producing strain of the receptor for the colicin, strains without receptor producing much larger zones (Fig. 2.1). They suggest that if receptor is present, then the colicin is released in association with it. It would seem possible that the association of some colicins with O antigen might be due to their specific absorption to receptor sites on this particular molecule. Nonetheless, it should be noted that colicins generally diffuse from a streak of growth on agar at a rate characteristic of the colicin type [LIKHODED, 1963 (2)], suggesting that at least the fastest moving component has a molecular weight characteristic of the type.

## Pyocins A2-P1-III and A3-P1-III

The cell wall of *Pseudomonas aeruginosa* P1-III has been the subject of intensive study by HOMMA and his colleagues (HOMMA, HAMAMURA, NAOI and EGAMI, 1958; HOMMA and SUZUKI, 1964, 1966; HOMMA, SUZUKI and ITO, 1963). The strain is pyocinogenic and it has been shown that there is some pyocin activity associated with the purified O antigen complex.

Using Goebel's "lower phase solvent" they were able to obtain from this complex a protein with reasonable pyocin activity (HOMMA and SUZUKI, 1964) although still much less active than purified colicin preparations. The protein appeared to be homogeneous on electrophoretic and ultracentrifugal analyses, and has a sedimentation coefficient of 1.15 S. The molecular weight is estimated by the Archibald method of calculation to be 42,000. This protein is known as pyocin A2 and would seem to be a protein associated with the O antigen of the cell wall. It is of interest that the protein and the lipopolysaccharide component can be made to reassociate and that under these conditions the pyocin activity is greatly reduced. Thus both in the original and in the reconstituted complex, the pyocin activity of the protein is masked.

The same group (HOMMA, GOTO and SHIONOYA, 1967; HOMMA and SHIONOYA, 1967; HOMMA and SUZUKI, 1966) have found another protein termed pyocin A3, which is purified from the cell cytoplasm. It has a molecular weight of 23,000 and also differs from A2 in its electrophoretic mobility. On the basis of activity spectra and cross-neutralisation with antisera, it seems probable that the two pyocins owe their activity to the same active site.

## Bacteriocins of Gram-positive Bacteria

### Megacin A-216

Megacin A is a bacteriocin produced by a gram-positive bacterium of the species *Bacillus megaterium*. It has been purified and characterised by HOLLAND (1961). The purified megacin was homogeneous when examined by analytical ultracentrifugation or by electrophoresis. It is a protein and, as might perhaps be expected of gram-positive bacteria, not associated with lipid or lipopolysaccharide components. It is not known whether this megacin has any association with the cell wall, as no studies on cell wall components were carried out. The purified megacin had a sedimentation coefficient of 4.3 S, and a molecular weight of 51,000 was calculated. This megacin is reported to be unaffected by all enzymes tested, including pepsin, trypsin and chymotrypsin (HOLLAND, 1961), although IVANOVICS, NAGY and ALFOLDI (1959) find their preparations to be inactivated by chymotrypsin.

OZAKI, HIGASHI, SAITO, AN and AMANO (1966) showed that megacin A has phospholipase A activity. They demonstrated the formation of lysolecithin from egg yolk lecithin by purified megacin A and also found that in the presence of lecithin, megacin A would cause hemolysis. The preparation of megacin A used gave a single protein peak on sucrose density gradient centrifugation and both bactericidal and hemolytic activities occurred as single peaks at the same place. Likewise on immuno-electrophoresis only a single precipitation band was observed for the purest material and both biological activities could be observed only up to this band. Thus it seems established that one and the same protein has both megacin A and phospholipase A activity.

### Megacin KP-337

This megacin has been studied by DURNER and MACH (1966). It was heat labile, 15 min at 60° almost completely destroying the activity. It was not dialysable and, contrary to previous reports (MARJAI and IVANOVICS, 1962), was not sedimented by 280,000 g for 1 h, indicating that it was not present as a large complex or particle.

### Staphylococcin 1262

An extract containing staphylococcin from *Staphylococcus* 1262 was studied by LACHOWICZ and WALCZAK (1966), who found that on Sephadex G-50 or G-75, 2 peaks of activity were obtained. They estimated from this that the two components have molecular weights of 30,000 and 9,700 and it was pointed out that one might be a trimer of the other.

### Enterococcin E1

An enterococcin produced by *Streptococcus faecium* E1 and studied by BRANDIS, BRANDIS and VAN DE LOO (1965) was partially purified and could be separated into two components by electrophoresis or by Sephadex G-75 chromatography; from the latter experiment they estimated the molecular weights of the components to be 50,000 and 100,000. The enterococcin of this strain, like those of most other *S. faecium* strains tested, was unaffected by trypsin or a bacterial protease. However, the active fractions obtained by sephadex chromatography were associated with protein peaks and analysis of the materials obtained showed the presence of many amino acids. The nature of the enterococcins is not yet known but they may well be proteins.

# Defective Bacteriophages

Defective bacteriophages are those unable to complete a full growth cycle for one reason or other and defective mutants can readily be isolated from phages such as λ (STENT, 1963). Bacteria lysogenic for defective phages may or may not lyse after induction, depending on which phage property has been lost. Some are unable to complete maturation while others lyse and release phage which for one reason or other cannot infect bacteria.

In some cases phage ghosts, consisting of protein coat only, can kill bacteria by adsorption (HERRIOTT and BARLOW, 1957). It is obvious that a defective phage, able to kill but not infect a bacterium, and released on induction of a suitable lysogenic strain, would closely mimic the biological properties of bacteriocins. It is not surprising then that some of the substances first described as bacteriocins have since been shown to be defective bacteriophages.

The "colicin" produced by *E. coli* 15 (MUKAI, 1960; RYAN, FRIED and MUKAI, 1955) has been shown to be a defective phage (ENDO, AYABE, AMANO and TAKEYA, 1965; FRAMPTON and BRINKLEY, 1965; MENNIGMAN, 1965; SANDOVAL, REILLY and TANDLER, 1965). The phage-like structure has a hexagonal head of 600—800 Å across the edges, and a tail 1,300—1,500 Å long and 200—300 Å in diameter. The biological activity is associated with this large structure as demonstrated by Sephadex G200 chromatography or centrifugation.

Several "pyocins" have been shown to have a phage-like structure. The first to be described was that produced in *Pseudomonas aeruginosa* R and studied by a group at Tokyo (ISHII, NISHI and EGAMI, 1965; KAGEYAMA, 1964; KAGEYAMA and EGAMI, 1962). The main structure observed by electron microscopy is very similar to the tail of a T-even bacteriophage, being a double, hollow cylinder 1,200 Å long and 150 Å in diameter. Other fragments of a phage were also present. The preparation contained only protein and no nucleic acid. It thus seems clear that this pyocin, in contrast to pyocins A2 and A3 of Homma's group, is in reality a defective phage.

The strain P1-III of *Pseudomonas aeruginosa*, which produces the protein pyocin A2, also produces a phage-like material lacking nucleic acids, and known as pyocin $A1_{mc}$ (HOMMA and SHIONOYA, 1967). Using antisera they were able to demonstrate cross-reaction between the pyocin $A1_{mc}$ and similar materials produced by other strains, and with some of the bacteriophages also produced by strain P1-III. There was no cross-reaction with the pycoins A2 and A3 discussed above. Other "pyocins" have been shown to have a similar structure (BRADLEY, 1967; HIGERD, BAECHLER and BERK, 1967; ONISHI, 1969; HOMMA, SHIONOYA, MEGURO and TANABE, 1967; TAKAYA, MINAMISHIMA, OHNISHI and AMAKO, 1969) and include one obtained from *P. aeruginosa* C. 10 (BRADLEY, 1967). However, since *P. aeruginosa* strains frequently produce more than one pyocin, it cannot be assumed that it is the same as that used in the experiments of JACOB (1954) discussed in Chap. 6.

Many other bacteria produce phage-like particles [BRADLEY, 1967; BRADLEY and DEWAR, 1966; HAMON and PERON, 1966 (7), 1967 (1); JAYAWARDENE and FARKAS-HIMSLEY, 1968; TAUBENECK, 1967; VIEU, CROISSANT and HAMON, 1967] and some are bactericidal. However, it is not yet clear how many of the previously recognised bacteriocins are phage-like in structure, nor indeed is it always certain that the phage-like structures observed are the active agents.

We should also mention here three defective bacteriophages of *Bacillus subtilis* studied by SUBBAIAH, GOLDTHWAITE and MARMUR (1965) which were recognised as having properties of bacteriocins while morphologically clearly identifiable as bacteriophages. They contain only host DNA and are detected by their ability to kill sensitive cells; one of them kills by inhibiting DNA and protein synthesis. Similar phenomena are likely to be found in other bacteria, and "bacteriocins" identified only by their biological rather than physicochemical properties may well be defective bacteriophages.

## Other Observations on Chemistry of Bacteriocins

There have been many observations made in the literature on the nature of bacteriocins based, not on the examination of purified material, but on examination of the effect of various specific agents on the biological activity of bacteriocin solutions.

Some colicins have been reported to be thermostable by FREDERICQ [1963 (1)], and LIKHODED [1963 (2)] found colicins of type D (5 studied), G (10) and V (5) to be at least partially stable to 100° whereas types I (5 studied) and B (5) were destroyed by 30 min at 60°. Diffusion rates through agar from a chloroform-killed streak of growth were again homogeneous within each type, colicin V diffusing faster than B, D, I and G in that order. The colicins of types I, B and V were sensitive to trypsin whereas under the same conditions colicins of type D were almost completely resistant. The 10 colicins of type E studied were variable in all these properties.

Of particular significance here is a comparative survey [HAMON and PERON, 1965 (4); HAMON, MARESZ and PERON, 1966] of many colicins and other bacteriocins, being an extension of earlier systematic observations by FREDERICQ. They divided the colicins into two groups, the first comprising colicins A, D, K, N, P and V, which in general are resistant to urea, are heat resistant, and resist the action of other agents such as ascorbic acid in the presence of cupric ions. The other group comprises colicins B, C, E, H, I and L, which are sensitive to these agents. Colicin E is also sensitive to chloroform. In addition to colicins they examined a range of other bacteriocins, in particular those produced by other strains from the family Enterobacteriaceae, and found that almost all of them were sensitive to the reagents listed above, resembling in this colicins B, C, E, H, I and L. The colicins of the first group, when purified, are often associated with O antigen whereas colicins of type E and I in the second group are not. This difference in association with other wall components may perhaps account for the differences in heat stability, etc.

Many other studies have included data on the sensitivity of bacteriocins to particular agents; we cannot list them all here, but will single out some of the more interesting. In addition to the colicins referred to above, some enterococcins, staphylococcins, lactocins and meningocins are heat stable [BARROW, 1963 (2); DE KLERK and COETZEE, 1961; KINGSBURY, 1966; LEONOVA, 1968]. A high proportion of the bacteriocins tested have been found to be inactivated by proteolytic enzymes, including many from both gram-positive and gram-negative bacteria. However, there are also many reports of bacteriocins which are unaffected by these enzymes, including, according to one account (HOLLAND, 1961), megacin A, for which a proteinaceous nature is well established. Most colicins are inactivated although the sensitivity varies considerably. Colicins C, D and S4 have been reported to be resistant to the *Proteus* enzyme (FREDERICQ, 1948), although HAMON *et al.* (1966) found all colicins to be sensitive to trypsin and chymotrypsin. However, not all colicins of a given type would necessarily be the same in this respect. Most other bacteriocins of gram-negative bacteria are sensitive, an exception being one of the two groups of marcescins (see Appendix). Bacteriocins of gram-positive bacteria are more commonly resistant to proteolytic enzymes but meningocins (KINGSBURY, 1966) and some enterococcins (BRANDIS *et al.*, 1965; BROCK, PEACHER and PIERSON, 1963) are sensitive.

Most bacteriocins for which we have any information are also nondialysable, and this in conjunction with the frequent occurrence of sensitivity to proteolytic enzymes serves to differentiate the bacteriocins as a group from other antibiotics, which in general are of low molecular weight and not proteinaceous although there are of course several important polypeptide antibiotics.

## Conclusions

The "bacteriocins" which have been purified fall into two classes: those which resemble bacteriophages or fragments of them, and those which are either simple proteins or proteins associated with cell wall components. Only the latter group are discussed in detail in this book as bacteriocins, and there is no evidence that they are in any way chemically related to bacteriophage proteins such as tail fibre proteins. Further discussion of the relationship between bacteriocins and bacteriophages must wait until Chap. 7.

Those bacteriocins which are simple proteins may be associated with cell wall components, and many colicins seem to be of this type. Others appear to be enzymes which cause cell lysis.

# Chapter 3

# Inheritance of Bacteriocinogeny

## Introduction to Bacterial Genetics

In this chapter we shall investigate the inheritance of bacteriocinogeny, although the available data will restrict us almost entirely to a consideration of colicinogeny. We will first very briefly survey the major characteristics of inheritance in bacteria; those who would like a full account are referred to "The Genetics of Bacteria" by Hayes.

At least some, and perhaps most bacterial species can undergo genetic recombination, but the mechanisms which have been evolved in the bacteria differ considerably from those of higher organisms, and are all independent of reproduction, which occurs by binary fission only.

In almost all strains of bacteria known to undergo genetic recombination, it is a rare event. Thus, if one mixes cultures of two strains, which differ in certain characters, then one may observe only one recombinant possessing a combination of parental characters for every $10^6$ or so bacteria in the initial cultures. For this reason students of bacterial genetics depend on selective techniques to isolate the recombinants for further study.

Take as an example *E. coli* K12, the strain in which bacterial recombination was discovered by LEDERBERG and TATUM (1946). It can grow on minimal medium which contains glucose and ammonium ions as sole carbon and nitrogen sources. Many mutant strains have been derived which cannot synthesise, and hence require the addition of a specific growth factor such as an amino acid or a purine base. If a mixture of two such cultures, each requiring a different growth factor, say methionine and proline respectively, is plated on minimal agar then any cell which has inherited the ability to synthesise methionine from the one strain and proline from the other will grow on the surface of the agar into a visible colony. Such colonies are indeed observed and demonstrate the existence of genetic recombination.

How do these recombinants arise? Recombinants have been observed to result from crosses in many bacterial species but in all cases they arise after the transfer of genetic material in the form of DNA from one cell, the donor, to another, the recipient. In some species this transfer is accomplished by a mechanism known as *transformation*, whereby free molecules of DNA, derived by either natural or artificial lysis of the donor cells, are taken up by the recipient cells. In other instances the donor and recipient cells *conjugate* and DNA is transmitted directly to the recipient across a conjugation bridge. The third mechanism is known as transduction and involves bacteriophages. In some strains of bacteriophage, a proportion contain DNA derived from the bacterial cell on which they were propogated. When such a phage injects its DNA into the next host, in addition to any phage DNA it also injects into that bacterium the bacterial DNA it is carrying; once again we have a distinction between donor and recipient.

Regardless of which of the three mechanisms operated, the recipient now contains its own complete genome and a portion of that of the donor. By means which are not yet fully understood, part of the donor fragment is incorporated into the recipient's genome, replacing the homologous portion of its own. We now have a recombinant bacterium with some of its genetic information derived from each of two parents.

The genome of bacteria has been shown to consist essentially of a single major chromosome which is a long strand of DNA, in *E. coli* 1,400 $\mu$ long (CAIRNS, 1962), consisting of a single Watson-Crick double helix. Additionally there may be relatively small genetic elements determining particular characteristics, such as fertility, or colicinogeny. We must now examine in some detail the fertility system of *E. coli*, which is determined by genetic elements similar to those determining colicinogeny.

### The Fertility System of E. coli K12

Much of the work on genetics of colicinogeny has been done using *E. coli* K12. Many mutants of this strain have been obtained and crosses between these mutants occur freely by conjugation (see HAYES, 1968 for references).

The ability of *E. coli* K12 to act as a donor of genetic material is determined by the presence of a small supernumerary chromosome, usually called a sex-factor and which we shall refer to as an epichromosome. Like the main chromosome the sex-factor is also circular. This epichromosome is not an essential component of the genome as it is sometimes lost spontaneously, perhaps on division, giving rise to a clone of cells unable to act as donors in conjugation.

Cells carrying the sex-factor are known as F+ and those lacking it as F−. Two F− bacteria cannot conjugate but F+ and F− cells conjugate very readily and the main chromosome may be transferred, albeit with a low probability, from the F+ to the F− cell. There is, however, a high probability that on conjugation a copy of the sex-factor will transfer to the F− cell, rendering it stably F+ and having the same properties of fertility as the original F+ cell. This transfer of the sex-factor is thought to occur linearly and to involve replication of the DNA, one of the daughter molecules being retained in the donor cell and the other passed into the F− cell. This special replication associated with transfer presumably begins at the same point on the circular sex-factor as normal replication and this point is perhaps permanently attached to the cell membrane.

The small sex-factor occasionally becomes interpolated into the main bacterial chromosome and such an association may be stable and give rise to a high frequency recombining strain (Hfr). The characteristics of Hfr strains show that the properties of the two original chromosomes are combined in one. In particular, on conjugation the major chromosomal portion is transferred with the smaller sex-factor at a high probability; hence the name Hfr. In this context we can think of the large bacterial chromosome being integrated into the smaller one and being transferred by it, the initiation point for transfer and associated replication being on the sex-factor but transfer once initiated continuing along the whole DNA sequence.

The chromosomal region transferred first is known as the origin of that particular Hfr strain and is transferred with high probability once conjugation occurs. However, transfer of the whole chromosome takes about 90 min and is not usually completed before conjugation is disrupted. Thus the terminal loci are transferred with a relatively low probability.

Since the point of integration of the sex-factor varies from one Hfr strain to another, the loci transferred early in conjugation will also vary with the Hfr strain involved. This can be demonstrated if conjugation is interrupted soon after it starts, when only loci which are transferred early will have entered the recipient and be able to be incorporated in recombinants.

Hfr cells are present in any culture of an F+ strain but it is not yet known if the limited fertility of F+ strains is due only to the low proportion of Hfr cells.

Hfr cells can occasionally revert to F+ cells, the sex-factor leaving the main chromosome. The sex-factor thus formed may also include some of the bacterial chromosome adjacent to the point of the original insertion. Thus an Hfr with the sex-factor inserted near the Lac genes (genes determining ability to ferment lactose) will occasionally generate F+ clones in which these genes are included in the sex-factor. This new epichromosome is called an F′ (F prime) factor and in this particular case an F′ Lac factor. A strain carrying F′ Lac will be a donor and transfer the Lac genes with high probability on conjugation. The Lac region of the chromosome has, in effect, been translocated to the epichromosome.

### The Sex Pilus

All three types of donor cells, F+, F′ and Hfr have a surface structure which is involved in conjugation and known as the F-type sex pilus; it is about 95 Å in diameter and up to 25 $\mu$ in length (BRINTON, 1965; LAWN, 1966). Certain bacteriophages, such as MS2, are specifically adsorbed to this structure and are able to infect only F+, F′ or Hfr strains of *E. coli* K12 or other strains with a similar pilus. They are known as male specific bacteriophages and are useful for recognition of the F-type pilus. During conjugation the pilus perhaps forms a bridge between the two conjugating cells.

### The Epichromosomes

Small chromosomes, similar to the sex-factor, are of considerable importance in bacterial genetics and we will find that some, known as C-factors, are involved in the inheritance of colicinogeny. The word 'episome' was coined in 1958 by JACOB and WOLLMAN to include those genetic elements which could exist either free or attached to the chromosome. The best-known are temperate phages and the sex-factors we have discussed already. Since then many other genetic elements have been discovered which resemble the sex-factor very closely, but it has not often been easy to demonstrate their ability to be integrated into the main chromosome. The word episome admirably emphasised the importance of reversible integration with the main chromosome shown by some genetic elements. Unfortunately such reversible integration has now proved unsatisfactory as part of a definition because of the difficulty of demonstrating it, except in certain special cases where integration gives new properties, as in Hfr strains. For this reason the sex-factors, together with

C-factors and other similar genetic elements, will be referred to as epichromosomes in this book. The term plasmid is also used by some authorities to encompass these various genetic elements; this term was introduced by LEDERBERG (1952) for genetic entities involved in extrachromosomal inheritance, and was at first used mostly for cytoplasmic factors of eucaryotes. It does not seem appropriate for the bacterial elements we are discussing which seem to be small chromosomes, rather than extrachromosomal. The word plasmid has also sometimes been used only for those factors not shown to integrate with the main chromosome (and hence excluded from the class of episomes).

The word epichromosome has long been used to describe the small chromosomes observed in certain invertebrate species, which may or may not be present in a given nucleus and are readily lost during cell division. Notwithstanding the differences between the chromosomes of invertebrates and bacteria, the epichromosomes of the two groups seem to bear a similar relationship to the respective major chromosomes.

Epichromosomes have been discovered in strains of several species of bacteria, in particular among species of the family Enterobacteriaceae. Each epichromosome is recognised because it (a) carries certain genetic information which has an observable effect on the cells in which it is present and (b) can be readily transferred by cell contact to other cells where the carried genes produce the same effect. An epichromosome, like the main chromosome of a bacterium, is replicated in phase with cell division, and its replication must be subject to regulation to ensure this. Unlike the main chromosome of *E. coli* K12, the epichromosomes, or at least any which we can hope to detect, can also, in some way, organise their own transfer from one cell to another during conjugation. In fact, conjugation itself seems to be mediated by the products of genes usually present on epichromosomes.

Some epichromosomes, like the sex-factor of *E. coli* K12, are able to confer a more general fertility upon a bacterium, promoting the transfer of the main chromosome into other bacteria during conjugation. Other epichromosomes are able to organise their own transfer only in the presence of an epichromosome of the former type, perhaps because they themselves do not carry all the genes necessary for the formation of functional conjugation tubes.

Some epichromosomes have been discovered during studies of fertile *E. coli* strains, and in fact the first evidence for an association between colicinogeny and fertility came from some experiments on fertility in *E. coli* (FURNESS and ROWLEY, 1957). The majority of epichromosomes have been discovered because, like the F′ strains of K12, they carry genes which are readily detectable upon transfer to another bacterial strain. Epichromosomes which have so far been detected in wild strains usually carry genes which determine resistance to one or more antibacterial drugs, or determine the production of a colicin. In the former case they are referred to as R-factors (Resistance factors), and in the latter as C-factors (Colicin factors).

# C-Factors

## The Epichromosome E1-K30

The first analysis of the genetic basis of colicinogeny was carried out by FREDERICQ and BETZ-BARREAU [1953 (1, 2, 3)]. They tested 12 colicinogenic strains to see if they would cross with an F⁻ strain of *E. coli* K12. Only one of the crosses gave any recombinants, that using *E. coli* K30 as the colicinogenic strain; this gave only seven recombinants which were poor growers on minimal medium, and in fact soon lost their ability to grow at all without the normal supplements of the K12 parent. An examination of their genetic characters showed them to resemble the K12 F⁻ parent in almost all respects, but 6 of them produced a colicin of type E. *E. coli* K30 itself produces colicins of types E and V. Thus, although the cross K30 × K12 F⁻ was only very slightly fertile, and the recombinants were unstable and soon reverted to the K12 parental type in most characters, the colicinogenic property was transferred from K30 with high probability to these rare recombinants and was then stably inherited. This suggested immediately that colicinogeny was being inherited in a different way to the other genetic characteristics.

Table 3.1. *The effect of the polarity of a cross on the inheritance of colicinogeny*
Auxotrophic or streptomycin-resistant derivatives of *E. coli* K12 were used in F⁺ × F⁻ crosses in which either parent was colicinogenic for colicin E1-K30. Details are in the text and are taken from FREDERICQ and BETZ-BARREAU [1953 (3)].

| Colicinogenic status of parents | | Markers selected from each parent | | % of recombinants colicinogenic |
|---|---|---|---|---|
| F⁺ | F⁻ | F⁺ | F⁻ | |
| col⁻ | col⁺ | Thr Leu Thi | Met | 100 |
| col⁺ | col⁺ | Thr Leu Thi | Str | 100 |
| col⁺ | col⁻ | Thr Leu Thi | Str | 34—73 |
| col⁺ | col⁻ | Met | Str | 40—73 |

Further crosses confirmed this suggestion. Thus, in a cross using a K12 F⁺ strain and a K12 (E1-K30) F⁻ strain obtained from the first cross, 100% of the recombinants were colicinogenic, although the segregation of the other genetic markers was identical to that in the cross K12 F⁺ × K12 F⁻. These recombinants were F⁺ and when crossed with an F⁻ strain 34—73% of the recombinants were colicinogenic. These two crosses are summarised in Table 3.1, and it can be seen that colicinogeny is not behaving like a normal genetic marker, for in that case we would expect the proportion of recombinants receiving a donor allele to be the same whichever allele it was.

On the contrary, colicinogeny is being inherited in these crosses as though it were carried by an additional genetic element or epichromosome, non-colicinogeny never being transferred from the donor while colicinogeny can be transferred with high frequency. FREDERICQ [1954 (1)] subsequently found that the transfer of colicinogeny from *E. coli* K30 to *E. coli* K12 occurred at high frequency independently of recombination for other markers. He thus inferred the existence of an epichromosome which confers the ability to make a colicin and is able to be transferred from cell to cell if a suitable conjugation system is present. It is transferred with high efficiency during conjugation from either an F⁺ or Hfr strain to an F⁻ strain but not in the opposite direction.

## Nomenclature of C-Factors

The epichromosome which determines the production of a colicin of a particular type may vary according to the strain from which it was originally derived and, as we have already observed, two colicins which are at one time thought to be indistinguishable may, on closer study, be shown to be significantly different. Thus a statement of the colicin produced is insufficient to define a C-factor, and each is named after its parent strain. Thus, *E. coli* K12 (K-K235) is *E. coli* K12 made colicinogenic for colicin K by a C-factor originally derived from *E. coli* K235, whereas K12 (K-K49) refers to a strain of *E. coli* K12 made colicinogenic for colicin K by a C-factor originally derived from *E. coli* K49.

## The Epichromosome I-P9

An epichromosome originally derived from *S. sonnei* P9 has been extensively studied and will serve as an example for us. *S. sonnei* P9 produces two colicins of types I and E2. After mixed growth overnight with *S. typhimurium* LT2, 50% of the LT2 were able to produce colicin I and 5% were able to produce both colicins (OZEKI, STOCKER and SMITH, 1962). These strains were indistinguishable in their biochemical characteristics from the original LT2 except for the production of colicins apparently identical to those of P9, both in their range of activity and in the appearance of their inhibition zones. Thus colicinogeny again appears to be determined by an epichromosome.

*S. typhimurium* LT2 (I-P9) was found to have acquired new properties of fertility, and in addition to the I-P9 factor, genetic markers on the main chromosome could now be transferred to other LT2 strains. The E1-K30, K-K49 and E2-P9 C-factors were also readily transferred by LT2 (I-P9), although rarely or not at all in the absence of the I-P9 C-factor. The I-P9 C-factor thus closely resembles the F-factor of *E. coli* K12 in its ability to promote genetic transfer (OZEKI and HOWARTH, 1961).

The I-P9 C-factor can exist in *S. typhimurium* LT2 in two different states (OZEKI, STOCKER and SMITH, 1962; STOCKER, SMITH and OZEKI, 1963), the first being that which characterises a cell to which the I-P9 C-factor has only recently been transferred, such a cell being able to transmit the I-P9 C-factor and any other C-factors present, at high frequency to a recipient. After a few generations, a different, more stable state is set up and further transfer is much less likely. Thus, when LT2 (I-P9) and LT2 (col⁻) are mixed in the ratio of, say 1 to 20, a few of the colicinogenic cells will transfer the I-P9 factor, and produce cells which can transfer the I-P9 factor to other cells with high probability. The I-P9 factor then multiplies faster than the cells and rapidly spreads throughout the population. Such a culture, for a brief time after the completion of this epidemic spread of the I-P9 factor, is termed an HFC culture (High frequency of colicinogeny transfer). An HFC culture will also give genetic recombinants at higher frequency than an LFC (Low frequency of colicinogeny transfer) culture (Table 3.2). It is interesting to note that if the intermediate culture (that used to allow the epidemic spread of I-P9 to make an HFC culture) carries the E1-K30 C-factor, then the frequency of recombination is even higher.

Table 3.2. *Recombination due to the I-P9 C-factor in the LFC and HFC states*

| Bacterial strain used | State of I-P9 factor in donor strain | Intermediate strain | Frequency of recombination |
|---|---|---|---|
| *S. typhimurium* LT2 | LFC | None | $0 \; (< 10^{-10})$ |
| | HFC | col⁻ | $10^{-8}$ |
| | HFC | (E1-K30) | $10^{-6}$ |
| *E. coli* K12 | LFC | none | $2 \times 10^{-9}$ |
| | HFC | col⁻ | $10^{-8}$ |
| | HFC | (E1-K30) | $10^{-8}$ |

From CLOWES (1964) summarizing data from other publications.

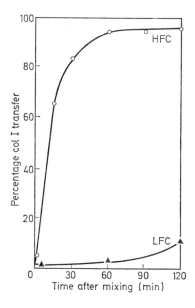

Fig. 3.1. *Low and High frequency transfer of the I-P9 C-factor.* LFC and HFC cultures of *E. coli* K12 were mixed with a recipient K12 strain at about $10^8$ organisms/ml each and incubated. At intervals samples were plated on streptomycin agar (only the recipient was resistant), the resulting colonies tested for colicin production and the percentage of the recipients which had become colicinogenic recorded. [From MONK and CLOWES, 1964 (1)]

CLOWES (1961) and MONK and CLOWES [1964 (1)] showed that I-P9 transferred to a K12 F⁻ strain conferred on this strain the same properties of fertility that it did on *S. typhimurium* LT2 (Fig. 3.1), and CLOWES (1964) has also summarised our understanding of the properties of I-P9 in both strains. The frequency with which each class of recombinants occurs using HFC cultures of *E. coli* closely resembles that produced by F⁺ strains. Although I-P9 is very similar in many ways to the K12 sex-factor there are points of difference, some of them obviously significant; for instance, the apparent non-existence of Hfr strains in I-P9 cultures (MEYNELL and EDWARDS, 1969).

Bacteria carrying I-P9 also have the ability to produce a sex pilus of different specificity to the F-type pilus (LAWN, MEYNELL, MEYNELL and DATTA, 1967; MEYNELL and LAWN, 1967). The pilus, which is known as an I-type sex pilus, or I pilus, has different antigenic specificity to the F-type pilus and confers susceptibility to a different group of male specific phages. The dimensions of the I-type pilus are also different. Pili are not present on many cells in an LFC culture but are present on most cells of an HFC culture of LT2 (I-P9).

The I-type and F-type pili probably serve the same role in conjugation, and are thought to act as the bridge through which the DNA passes from one cell to another.

### The Epichromosomes VB-K260 and VI-K94

Several C-factors have been shown to act as promoters in addition to I-P9, but so far only two have been shown to enter the Hfr state. FREDERICQ [1963 (2), 1969] showed that the VB-K260 C-factor could promote transfer of chromosomal genes from *E. coli* K12. One of the recombinants obtained using *E. coli* K12 (VB-K260) as a donor was found to be itself a donor of Hfr type, and to produce colicin B but not V. Xyl was the first locus transferred and Str the last by this particular strain. The colicinogeny B and the Hfr determinant were located near the terminal marker Str, all three properties being closely linked. It appears that the B-K260 epichromosome is very similar in properties to the sex-factor of *E. coli* K12 in that it can be integrated into the chromosome to give an Hfr and colicinogeny B is then transferred as the terminal marker.

KAHN (1968) isolated 18 independently derived Hfr strains from *E. coli* K12 (VI-K94). From three derivatives of *E. coli* K12 she obtained 13 Hfr strains, of which 12 were identical in origin and direction of transfer, closely resembling the Hfr strain derived by FREDERICQ from VB-K260. However, all of five Hfr strains obtained from a further derivative of *E. coli* K12 had unique combinations of origin and direction of transfer. She concluded that most derivatives of *E. coli* K12 had a chromosomal site (near Xyl) with considerable homology with a part of the VI-K94 epichromosome. The genes for colicinogeny were still functional in the Hfr strains and were apparently linked to the terminal markers on transfer, and on this basis would appear to be integrated into the main chromosome. However, some of these Hfr strains tended to lose their colicinogeny spontaneously while retaining their Hfr characteristics, and this is difficult to explain on the model of full integration of a single epichromosome into the main chromosome.

## The Inheritance of Other Bacteriocins

Studies on the genetics of bacteria are usually confined to the few strains in which recombination is known to occur and from which the mutants necessary for selecting recombinants have already been isolated. Thus our knowledge of inheritance of colicinogeny is virtually confined to those cases where the property can be transferred to suitable strains, either *E. coli* K12 or *S. typhimurium* LT2. The best statistical study is that of FREDERICQ [1956 (1)], who examined 314 colicinogenic strains of *E. coli* and related species to determine their ability to transfer the colicinogenic property to *E. coli* K12. Some of these 314 strains produced two or three colicins, not necessarily transferred together, and he had a total of 371 colicinogenic properties to study. The results are shown in Table 3.3, from which it can be seen that some colicinogenic properties, such as I, are frequently transmitted, whereas others are

Table 3.3. *Proportion of colicinogenic strains which can transfer the colicinogenic property to E. coli K12.* [*From* FREDERICQ, *1956 (1)*]

| Colicin produced by the original strains | Number of strains able to transfer | unable to transfer | Total |
|---|---|---|---|
| A | 0 | 7 | 7 |
| B | 17 | 52 | 69 |
| D | 0[b] | 7 | 7 |
| E | 16[a] | 55 | 71 |
| I | 21 | 8 | 29 |
| K | 4 | 15 | 19 |
| P | 0 | 48 | 48 |
| S4 | 0 | 12 | 12 |
| V | 2 | 104 | 106 |
| Others | 0[b] | 3 | 3 |
| Total | 60 | 311 | 371 |

[a] 8 of subtype E1, 2 of E2 and the others unclassified (FREDERICQ, personal communication).

[b] More recently the C-factor D-CA23 has been demonstrated by transfer to a *Providence* strain (COETZEE, 1964) and the C-factor G-CA46 by transfer to another *E. coli* strain (LORKIEWICZ *et al.*, 1965)

only rarely so and yet others, such as colicinogeny P, not at all. Thus FREDERICQ [1956 (1)] discovered 59 C-factors and one of them, VB-K260, was thought to confer two colicinogenies, thus accounting for the total of 60 in Table 3.3. The colicinogenic strains of K12 produced a colicin of the same type as the strain donating the C-factor [FREDERICQ, 1954 (1)]. Even when transfer does occur, the probability can vary considerably; in some cases all surviving cells of the K12 strain are colicinogenic, in others less than 0.001 %. However, since the recipient *E. coli* K12 is sensitive to the colicin, but is immune once it becomes colicinogenic, cells which do not receive the C-factor may be killed by colicin produced and so distort the percentages. The epidemic transfer of certain C-factors such as I-P9 will also increase the apparent frequency of transfer and probably accounts for the very high figures sometimes observed.

HAMON (1957) also studied the transfer of colicinogeny from one strain to another, and again found the new colicinogenic strain to produce a colicin identical to that of the previously colicinogenic strain. He also found that, as expected for epichromosomal inheritance, the new strain usually resembled the non-colicinogenic parent strain in all respects except colicinogeny and immunity to that colicin. In some instances, however, the strain also showed other new properties; in particular some *Salmonella* strains have a modified lysotype; i.e., their sensitivity to certain bacteriophages has changed. Perhaps the immunity conferred by the C-factor extends to bacteriophages, but this does not in any way invalidate the hypothesis that epichromosomal inheritance is involved.

Other studies have also demonstrated the existence of C-factors. Thus OZEKI *et al.* (1962) showed that 12 *S. typhimurium* strains producing colicin of type E2 could not transfer the property but all 12 could do so if made I-P9+, and LEWIS and STOCKER (1965) showed that of 7 *S. typhimurium* strains producing colicins of type E1, 5 could transfer the property to *S. typhimurium* LT2.

# Intergeneric Transfer

Transfer of colicinogeny has been shown to occur from *E. coli* to *Shigella* [FREDERICQ, 1954 (1)], *Salmonella* (HAMON, 1957), *Serratia* (AMATI and OZEKI, 1962), *Proteus* (DE WITT and HELINSKI, 1965), Providence strains (COETZEE, 1964), *Klebsiella* and *Paracoli* as well as to *E. coli*. However, only a few types of colicin are transferable as can be seen from Table 3.3. The bacteriocins of *Shigella* [FREDERICQ, 1954 (1)] have also been shown to be transferable by epichromosomes, and HAMON (1964) refers to unpublished data showing that bacteriocins produced by *Salmonella*, *Arizona* and *Serratia* can be transferrable. LORKIEWICZ, MACHIAZEK and NACHIEWICZ (1964) studied the ability of 9 *E. coli* strains to act as recipients for 3 colicinogenic strains; *E. coli* CA42 transferred E2-CA42 to one strain at a high frequency (75%) whereas the other 8 did not act as recipients for CA42. The same strain could also act as a recipient for I-CA53 and another untyped C-factor. Of the other 8 strains, 2 could act as recipients for I-CA53 and none for the untyped C-factor. Thus there is variation not only in donor ability but also in ability to act as recipient in specific crosses.

Epichromosomes determining pneumocin production in *Klebsiella* have been transferred both to *E. coli* strains and *Klebsiella pneumoniae* K9 (DURLAKOWA and LACHOWICZ, 1967; HAMON, MARESZ, HSI and PERON, 1967; HAMON, MARESZ, KAYSER and PERON, 1970; KAYSER, HAMON and LAMBLIN, 1971; STOUTHAMER and TIEZE, 1966). (See Appendix for discussion of pneumocins.) Some of these epichromosomes can transfer between strains of *E. coli* K12 (HAMON *et al.*, 1967), and some resemble Ia or Ib C-factors (HAMON *et al.*, 1970). All our positive knowledge of the genetic nature of bacteriocinogeny concerns members of the Enterobacteriaceae, and this may be because most of the others have yet to be studied. However, BEN-GURION and HERTMAN (1958) were unable to demonstrate the involvement of a transferable epichromosome with pesticin 1 synthesis, and HOLLAND and ROBERTS (1963) were equally unsuccessful with 7 megacinogenic strains.

## The Autonomous Nature of C-Factors

When it was realised that colicinogeny was frequently determined by epichromosomes, it was natural that attempts were made to demonstrate if, like the sex-factor and some bacteriophages, they could attach to the main chromosome. FREDERICQ and BETZ-BAREAU [1953 (2)] found that in any cross of K12 F+ (E1-K30) x F⁻ bacteria, the distribution of the colicinogenic property among the recombinants does not show any linkage to the other loci studied, and on this basis they considered that colicinogeny was inherited on an autonomous genetic element. However, on the basis of some experiments by ALFOLDI, JACOB, WOLLMAN and MAZÉ (1958), the E1-K30 factor was at one time thought to exist integrated into the chromosome and to be located between the Thi and Thr genes, and C-factors were one of the three different entities first grouped together as episomes by JACOB and WOLLMAN (1958). Since that time C-factors have often been referred to as episomes, with the implication that they do integrate with the chromosome, and perhaps we should examine some of the experiments demonstrating that they are not normally integrated into the chromosome. The V-K94, E2-K317, B-K98, E1-K30 and I-CA53 C-factors have been separately transferred into various Hfr strains with different origins of transfer, which have then been crossed with a non-colicinogenic F⁻ strain (NAGEL DE ZWAIG, ANTON and PUIG, 1962; NAGEL DE ZWAIG and PUIG, 1964; PUIG and NAGEL DE ZWAIG, 1964). They found that each C-factor was transferred at a characteristic frequency, and although this frequency varied considerably, for any one C-factor it was virtually independent of whether the colicinogenic strain was F+ or Hfr, or of the origin of the Hfr. They concluded that the C-factors had not integrated into the chromosome. CLOWES (1963) did similar work with the E1-K30 C-factor. He used various colicinogenic Hfr strains made E1-K30, and showed that, whatever the origin of the Hfr, E1-K30 was transferred at between 7 and 12 min; thus even E1-K30, which provided the original evidence for an episomal nature of C-factors, has not yet been demonstrated to integrate into the chromosome. The data of ALFOLDI et al. (1958), which were used as a basis for including C-factors in the category of episomes, have an explanation which, although very interesting, is quite unrelated to colicinogeny (BEN-GURION, 1963; CLOWES, 1963; NAGEL DE ZWAIG and PUIG, 1964). However, although most C-factors are not known to integrate with the chromosome, we must remember that the F-factor was also not shown to integrate until after several years of work on *E. coli* K12, and then only because clones in which integration had occurred had the distinctive property of Hfr donor ability. It is surely not a coincidence that the VB-K260 and VI-K94 C-factors, which are the only ones yet shown to integrate into the chromosome, are also promoters and give rise to "integrated" clones, initially detected because of their Hfr properties.

## The Types of C-Factor

When we first discussed epichromosomes, we said that in order to be easily detected as such, they must be able to transfer from one cell to another. However, some C-factors cannot transfer to another cell unless another fully competent epichromosome is present in the donor cell.

It seems that all the essential genetic properties for acting as a donor in con-jugation are carried by only a few epichromosomes, and these can transfer without help. Those in this category known to date are given in Table 3.4. Some are also able to promote chromosome transfer and transfer of other epichromosomes in addi-tion to transfer of their own genetic material, and these are listed separately in the same table. In addition, B-K77 and E1-K30 (OZEKI et al., 1962; PUIG and NAGEL DE ZWAIG, 1964) and also C-factors for colicin E1 from five strains of S. typhimurium (LEWIS and STOCKER, 1965) have been recorded as able to transfer from one cell to another without other helper epichromosomes, but have not yet been shown to promote transfer of chromosome or other epichromosomes. The E1-K30 C-factor will transfer between S. typhimurium LT2 strains, but not between E. coli K12 strains. The basis for this difference between two host strains is not known.

The other group of C-factors are those which cannot transfer from one cell to another without a helper epichromosome also being present in the donor. They are also listed in Table 3.4. The C-factors in this group are capable of transferring from one cell to another if the donor carries either I-P9 or the K12 F-factor, these being the only two epichromosomes which have been studied in respect of their helper ability for these particular C-factors. The majority of the C-factors whose existence has been demonstrated have not been studied in any detail, and it is not possible to place them in either of the two categories mentioned above; it remains possible that some of them will fall into the remaining possible category, not being transmitted even in the presence of either of the two known helper epichromosomes.

It will have been noted earlier that only 58 out of 314 colicinogenic strains could transmit their colicinogeny [FREDERICQ, 1956 (1)]. One possible reason for non-transmittance could be that the colicinogenic strain lacks the ability to conjugate, and OZEKI et al. (1962) have shown that some S. typhimurium strains producing colicin E2 are unable to transfer this ability, but can transfer it if the I-P9 epichromosome is first transferred to them. The I-P9 epichromosome presumably provided the ability to conjugate, making it possible to demonstrate the epichromosomal inheritance of these E2 colicinogenic properties. However, inability to conjugate is not the only reason for nontransfer of colicinogeny, as included in FREDERICQ's 58 strains are 16 that produce two colicins but transfer only one of them. In the case of strains which produce two colicins, and transfer both properties, we can ask if they are both on the same epichromosome. S. sonnei P9 produces two colicins, and, as we saw earlier, the C-factors I-P9 and E2-P9 are readily separated. However, E. coli K260 producing colicins V and B normally transfers both together and both are thought to be on the one C-factor (FREDERICQ, 1969).

# Transduction of Colicinogeny

The colicinogenic property may be transduced by bacteriophage P22 in S. typhimurium LT2 and by phage P1 in E. coli K12, although this has only been demon-strated for some C-factors [FREDERICQ, 1958 (1), 1959]. Phage P22 could transduce an unspecified C-factor of type E2 and phage P1 could transduce two unspecified C-factors of types E1 and B. The frequency of transduction was low and it was thought that some at least of the negative results obtained for other C-factors might have

Table 3.4. *Properties of various C-factors*

| C-factor | Synonym | Pilus type | Self transferable | Promoter of other DNA | Integrates to give Hfr | Inducible | Effect of Rec gene on induction | Epidemic spread | Coexists with I-P9 | Prevents I-P9 spread | FDR mutants isolated | Prevents E1 a spread | Coexists with F | Represses F fertility | Curable by acridine orange | Curable by thymine deprivation | References |
|---|---|---|---|---|---|---|---|---|---|---|---|---|---|---|---|---|---|
| VI-K94 | V2 | F | + | + | + | − | − | | | | | | − | | ± | + | 1, 9, 10, 11, 12, 13, 17, 21, 22, 24 |
| V-K30 | V3 | F | + | + | + | | | | | | | | − | ± | ± | + | 13, 17 |
| VB-K260 | | F | + | + | | | | | | | | | | | | | 6, 7, 8 |
| B-K77 | B2 | F | + | | | + | | + | | | | | + | ± | | | 3, 13, 17 |
| B-K166 | B3 | F | + | | | + | | + | | | + | | + | − | | | 3, 13 |
| B-CA98 | B1 | F | + | | | | + | + | | | + | | + | + | | | 3, 28 |
| B-K98 | B4 | F | + | − | | | + | − | | | | | + | ± | | | 3, 28 |
| E1-K30 | | F | − | − | | | | | | − | | | | − | − | + | 3, 4, 9, 11, 14, 17, 23, 29 |
| E2-P9 | | | | | | | | | | | | | | | | | 9, 14, 27 |
| E2-K317 | | | | | | + | | | | + | | | | | − | | 14, 24 |
| A-CA31 | | | | | | | | | | | | | | | | | 16, |
| Ia-CA53 | | I | | | | − | | | | | | + | | | + | | 13, 15, 16, 18, 24 |
| Ia-CT2 | | I | | | | | | | | | + | | | + | | | 13. |
| Ib-P9 | | I | + | + | | | | + | | | | + | + | | − | + | 2, 4, 5, 13, 17, 18, 19, 20, 25, 26, 27 |
| Ib-CT4 | | I | + | | | | | + | + | + | + | | + | − | | | 2, 13, 26 |
| E1-16 | | I | + | | | | | + | + | + | + | | + | − | | | 5, 13, 14, 18 |
| E1-18 | | I | + | | | | | | | + | + | | | | | | 5, 13, 14, 18 |
| K-K49 | | | − | | | | | | | | | | | | | | 14, 27 |
| K-K235 | | | − | − | | | | | | | | | | | | | 14 |

*References:*

1. Caro and Schnos, 1966
2. Clowes, 1961
3. Clowes et al., 1969
4. Clowes et al., 1965
5. Edwards and Meynell, 1968
6. Frederico, 1963 (3)
7. Frederico, 1964
8. Frederico, 1969
9. Helinski and Herschman, 1967
10. Kahn, 1968
11. Kahn and Helinski, 1964
12. Kahn and Helinski, 1965
13. Lawn et al., 1967
14. Lewis and Stocker, 1965
15. Likhoded, 1963 (1)
16. Lorkiewicz et al., 1965
17. MacFarren and Clowes, 1967
18. Meynell and Lawn, 1967
19. Monk and Clowes, 1964 (1)
20. Monk and Clowes, 1964 (2)
21. Nagel de Zwaig, 1966
22. Nagel de Zwaig and Anton, 1964
23. Nagel de Zwaig and Puig, 1964
24. Nagel de Zwaig et al., 1962
25. Ohki and Ozeki, 1968
26. Ozeki and Howarth, 1961
27. Ozeki et al., 1962
28. Puig and Nagel de Zwaig, 1964
29. Smith et al., 1963

been due to the difficulty of detecting low-frequency transduction of colicinogeny; it is not certain that any C-factors are not transducible. No linkage was detected with other bacterial genes. OZEKI and STOCKER (1958) found that the C-factor E2-P9 could be transduced by P22. Transduction has made possible the production of E2-P9 strains which do not carry I-P9, such strains being difficult to prepare otherwise as E2-P9 is normally only transferred with I-P9 by conjugation.

## Physical Nature of C-Factors

C-factors, like other genetic elements, are thought to be composed of DNA. This is, of course, expected from our knowledge of the role of DNA in genetic systems and there is also good experimental evidence on this point for some C-factors. Part of this evidence involves radioactive labelling of the DNA of cells carrying a C-factor. The amount of DNA associated with a C-factor is then estimated by measuring the amount of DNA transferred during conjugation. Such experiments suggest that I-P9 contains $6 \times 10^4$ base pairs, E1-K30 $7 \times 10^4$ base pairs, and E2-P9 $3 \times 10^4$ base pairs (SILVER and OZEKI, 1962). An alternative technique after transfer of the $^{32}$P labelled epichromosome is to freeze the bacteria and allow decay of the $^{32}$P (DRISKELL-ZAMENHOF and ADELBERG, 1963). ROTH and HELINSKI (1967), BAZARAL and HELINSKI [1968 (1, 2)] and CLEWELL and HELINSKI (1970) have been able to examine the DNA of several C-factors directly by electronmicroscopy, or by sucrose density gradient centrifugation. They all consist of supercoiled or open circles of double stranded DNA. This confirmed that the C-factors were composed of DNA and was also the first direct demonstration of the circular form of an epichromosome. Those of E2-P9 and E3-CA38 had molecular weights of $5 \times 10^6$, corresponding to about $8 \times 10^3$ base pairs, while E1-K30 had a molecular weight of $4.2 \times 10^6$ corresponding to about $6.5 \times 10^3$ base pairs. The DNA of E1-K30 obtained after transfer to *Proteus mirabilis* also included supercoils of 2 and 3 times the basic size, but their significance is not understood [BAZARAL and HELINSKI, 1968 (2)]. The I-P9 C-factor was much larger with a molecular weight of $61.5 \times 10^6$.

Another colicin of type E1, produced by a Providence strain, has also been shown by electron microscopy to consist of circles of DNA of about $4.2 \times 10^6$ M.W., together with circles of 2 and 3 times this basic size (VAN RENSBURG and HUGO, 1969). The sizes obtained by these more direct means are lower than expected from the earlier results and show that C-factors consist of DNA of about 0.14 to 0.17% of the size of the main chromosome.

## Conclusions

In this chapter we have seen that colicinogeny has a marked tendency to be inherited on epichromosomes. In one study [FREDERICQ, 1956 (1)] 16% of colicinogenic properties studied were inherited in this way, and in fact, although as yet nothing is known of the location of the pertinent genes in the other 84% of the strains, it may indeed be that all are located on epichromosomes but that for technical reasons this cannot yet be demonstrated. We have already seen one example (OZEKI *et al.*, 1962) in which special techniques were required to demonstrate the presence of the epichromosome. Alternatively, it is possible that the property of colicinogeny may also frequently be inherited as a locus on the major chromosome, and it is even possible that some of the C-factors which have been demonstrated arose immediately prior to their transfer to the recipient by interaction of a sex-factor and the main chromosome. If such C-factors (equivalent to F′ factors) were readily transferred whereas the sex-factor concerned did not otherwise promote transfer of the main chromosome, then the technique of isolating recipient bacteria which are colicinogenic would naturally select out strains carrying these newly formed C-factors; we have no way of knowing if colicinogeny was inherited in the original wild type strain not on an epichromosome, but on the main chromosome.

The colicinogenic strains which are able to transmit a C-factor to either of the two strains widely used in these studies, *E. coli* K12 and *S. typhimurium* LT2, are not randomly distributed; colicinogeny E and in particular E1, B and I are frequently determined by C-factors although there are exceptions; K and V are rarely determined by demonstrable C-factors and only one instance has been reported of a C-factor for colicin G (LORKIEWICZ, DERYLO and FRELIK, 1965) while many colicin types have never been reported to be inherited on epichromosomes.

C-factors were shown to consist of DNA. In addition to carrying a structural gene for a colicin, they confer on a cell immunity to that colicin and also some properties concerned with conjugation and transfer of the C-factor to other cells. The C-factors are quite variable in their properties, as indeed are the R-factors, the other major known group of epichromosomes (see NOVICK, 1969, for review).

# Chapter 4

# The Properties of C-Factors

C-factors by their very nature must carry one or more structural genes for the colicin protein, and presumably carry genes involved in regulating colicin biosynthesis. They may also carry genes concerned with such functions as replication of the epichromosome itself and its transfer to other organisms. We do not know as yet what significance to attach to the observation that the genes for colicinogeny are often carried on an epichromosome; however, since the control of colicin synthesis seems to interact with the control of other C-factor properties, we will have to consider these other properties.

## Regulation of Colicin Synthesis and C-Factor Replication

### The Production of Bacteriocins

Very early in the study of colicins it was reported that their production did not occur under all growth conditions, thus some colicinogenic strains do not normally produce any colicin at all in broth although they do so when grown on the surface of an agar plate. Those who have been interested in the purification of colicins have often reported that certain components of the medium appear to be critical for production, although these requirements appear to differ from one example to another. The studies on the influence of environmental conditions on the synthesis of colicin K by *E. coli* K235 (GOEBEL, BARRY and SHEDLOVSKY, 1956; MATSUSHITA, Fox and GOEBEL, 1960) and of pesticin 1 synthesis (HERTMAN and BEN-GURION, 1958) were particularly detailed.

In an interesting study, FOULDS and SHEMIN (1969) showed that the apparent temperature dependence of production of a marcescin was in fact due to the lability of a bacterial protease at 39°. At lower temperatures the enzyme was active and destroyed any marcescin produced. This example should warn us that the effects of the environment on bacteriocin production may be indirect or complex.

### Induction of Colicin Production

Soon after the demonstration that lysogenic bacteria could be induced to produce mature bacteriophages, JACOB *et al.* (1952) showed that *E. coli* ML could be induced by ultraviolet irradiation to produce large amounts of its colicin ML-E1. This induction was similar in many respects to the induction of prophage. A culture of *E. coli* ML grown in a synthetic medium produced no detectable colicin until induced by irradiation, when colicin was detected intracellularly within 10 min and rose to 150 arbitrary units per ml. At 60 min, extracellular colicin was detected and massive cell lysis occurred at about 90 min, releasing the colicin (Fig. 4.1). The final amount of colicin released was about 200 arbitrary units per ml or 40 lethal units per bacterium. In broth a yield of 500 lethal units per bacterium can be obtained and a rather higher yield of about 5,000 lethal units per bacterium is obtained on irradiation of other strains, e.g. *E. coli* CA42 (colicin E2). (A lethal unit, the amount of colicin which can kill one sensitive bacterium, is discussed in Chap. 5.)

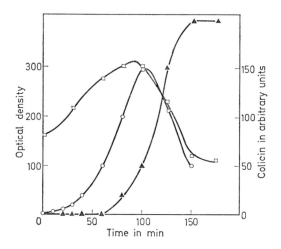

Fig. 4.1. *Production of colicin by irradiated bacteria. E. coli* ML grown in synthetic media was irradiated with ultraviolet light at time 0. At intervals a sample was removed and bacteria separated by centrifugation, broken with sand and the colicin activity associated both with culture supernatant and the broken bacteria was assayed. Optical density □ ——— □. Intracellular colicin O ——— O. Colicin in the culture medium ▲ ——— ▲. (From Jacob *et al.*, 1952)

*E. coli* ML has since been shown to be lysogenic [Fredericq, 1954 (2)] and part if not all of the inducibility may have been due to the presence of inducible prophage. It was also shown [Fredericq, 1954 (2)] that other colicinogenic strains were inducible and that upon transfer of the C-factor to *E. coli* K12 ($\lambda^-$) or *E. coli* B, the newly colicinogenic strains were also inducible even in the absence of any known prophage. However if one compares pairs of colicinogenic derivatives of *E. coli* K12 differing in whether or not they carry $\lambda$ prophage, then the kinetics of colicin induction are different [Fredericq, 1955; Hamon and Peron, 1962 (3, 4)]. In the non-lysogenic derivatives colicin is released continuously from the cells from about 10 min after induction whereas in strains which are $\lambda^+$ and also colicinogenic the colicin may be retained within the cell until phage "induced" lysis occurs with massive release of both the phage and colicin. This situation applies for E1-K30 and B-CA18 but in the case of K-K235 and E2-P9 the presence of the C-factor inhibits phage synthesis.

In addition to ultraviolet irradiation, other agents can also induce bacteriocin production, including mitomycin, peroxide, acridine orange, and N-methyl-N-nitrosoguanidine (DE WITT and HELINSKI, 1965; IIJIMA, 1962; OZAKI et al., 1966; SMARDA, 1962; SMARDA, KOUDELKA and KLEINWACHTER, 1964). Thymine deprivation can also induce bacteriocin production (SICARD and DEVORET, 1962). A series of anti-cancer agents have been tested, and many induced megacin synthesis by B. megaterium 216 (MARJAI and IVANOVICS, 1964).

In addition, some mutants of E. coli in which DNA synthesis is temperature sensitive, if also colicinogenic, are induced to synthesise colicin when DNA synthesis is arrested (KOHIYAMA and NOMURA, 1965).

Although most of the detailed studies on inducibility of bacteriocins concern colicins, many other examples have been reported; they include examples of caratovoricins, cloacins and pneumocins amongst the Enterobacteriaceae, fluocins and pyocins among the Pseudomonadaceae, and both pesticins. Amongst the gram-positive bacteria monocins, megacins and meningocins have been reported to be inducible. However, there have been no comprehensive surveys and frequently the data are not given; there seems little reason to attempt a comprehensive list.

### Lethal Biosynthesis of Bacteriocins

Colicinogenic bacteria, when plated on a lawn of an indicator strain, may give rise to a small zone of inhibition or "lacuna" [FREDERICQ, 1950 (2); GRATIA, 1932; OZEKI, STOCKER and MARGERIE, 1959). These lacunae are due to colicin released by single cells and the number of lacunae corresponds to about 0.1% of the bacterial culture if LT2 (E2-P9) is used, but to over 50% after induction of this strain by ultraviolet light (OZEKI et al., 1959). Since the lacunae do not contain a central colony, it was concluded that normal colicin production is due to occasional cells synthesizing a large amount of colicin and in the process dying. On induction the majority of cells are induced to synthesize colicin and die. OZEKI et al. (1959) used bacteria producing colicins of types B, E, I and K and they all produced lacunae on sensitive strains. They also observed colicin production by single bacteria using LT2 (E2-P9) after isolation by micromanipulation. These results confirmed that colicin production is by occasional lethal synthesis.

### The Induction of the E1-K30 C-Factor and Its Colicin

The similarities between the induction of colicin production and of lysogenic bacteria encouraged people to look at induced colicinogenic bacteria to see if the C-factor itself were replicating.

The first study was that of AMATI (1964), who used E. coli K12 derivatives carrying combinations of E1-K30, E2-P9 and I-P9. He found that in general after ultraviolet irradiation, colicinogenic strains which were induced to form colicin synthesized DNA at a faster rate while increasing in size more slowly. He took this to be evidence that the C-factors themselves replicated and, making certain assumptions, estimated the extent of such replication.

However, the situation only became clear when DE WITT and HELINSKI (1965) used equilibrium density gradient centrifugation to distinguish between the DNA of the epichromosome and the main chromosome. They transferred the E1-K30 C-factor to *Proteus mirabilis*. These cells then had, in addition to the *Proteus* DNA with a density of 1.700, a small amount (about 0.3% of the total) of DNA resembling that of *E. coli* in density, which they took to be the epichromosomal DNA.

On induction by mitomycin C, this epichromosomal DNA increased to somewhere between 0.92 and 9.9% of the total DNA depending on conditions and was correlated with an increase in colicin titre from virtually 0 to 10 or 300 units per ml respectively. The number of lacunae demonstrated after induction in different experiments was said to correlate with the other two measurements. These data suggest that on induction the number of C-factors increases from 3 to 30-fold and a certain amount of colicin is synthesised.

### The Role of C-Factor Proteins in Induction

It has been suggested that *de novo* protein synthesis is not required at least for the lethal effects of colicin induction [HERSCHMAN and HELINSKI, 1967 (1)]. However, the experiments involved the use of chloramphenicol, and it seems possible that if protein synthesis were necessary, it could have occurred after the dilution required to see if the cells were indeed viable. However, their second observation, that chloramphenicol prevented newly transferred E2-P9 from becoming inducible, suggests that a protein encoded by the C-factor itself is involved in induction.

### The Role of the E. coli Rec Gene in Induction

Mutants of *E. coli* unable to act as recipients in crosses are known as Rec⁻ (CLARK and MARGUILES, 1965). RecA⁻ mutants of *E. coli* K12, when lysogenic for λ, are not inducible by ultraviolet irradiation (FUERST and SIMINOVITCH, 1965) and it has been shown that in these mutants, the irradiation does not lead to inactivation of the repressor maintaining the prophage state (BEN-GURION, 1967; BROOKS and CLARK, 1967; HERTMAN and LURIA, 1967).

*E. coli* K12 (E1-K30) resembles *E. coli* (λ) in that induction (in this case of colicin synthesis) does not occur if the strain is RecA⁻ (HELINSKI and HERSHMAN, 1967). As yet it is not possible to define the role of the Rec gene in genetic recombination, or in the induction of lysogenic and colicinogenic bacteria.

### The Nature of E1-K30 Induction

Some lysogenic bacteriophages and some colicins are inducible, and this is one of several analogies between bacteriocins and bacteriophages. However, the significance of this analogy is not clear and it is in fact possible to see similar behaviour in the *E. coli* K12 main chromosome. Thus PRITCHARD and LARK (1964) have shown that when *E. coli* DNA resumes replication after being stopped by thymine deprivation, it not only continues at the point where it left off, but a new round of replication is started prematurely at the normal initiation point. It has also been shown to occur after ultraviolet irradiation (BILLEN, 1969; HEWITT and BILLEN, 1965). This premature initiation has no dramatic effects but clearly resembles the induction of $\lambda$ or E1-K30 replication and it seems possible that the replication of many different bacterial replicons (units of DNA replication) is inducible by certain treatments such as ultraviolet irradiation. To pursue this hypothesis further, it is possible that the repeated replication of epichromosomes after induction is due to their small size, many copies being synthesised before normal restraints can be reorganised, by which time they are unable to repress replication of the now large number of epichromosomes present.

The simultaneous induction of replication of the E1-K30 epichromosome and of the colicin it encodes, suggests that the two phenomena may be causally related. One possible explanation is that colicin synthesis is derepressed as a result of the rapid replication of the epichromosome. Other examples of escape synthesis, as it is known, are the Lac genes when present in the transducing phages P1dL (REVEL and LURIA, 1961) or $\phi$ 80dL (PFAHL, 1969) and the Gal genes when present in the transducing phage $\lambda$dG (BUTTIN, 1961; GERMAINE and ROGERS, 1970). In both cases induction of the prophage leads to rapid synthesis of proteins encoded by the DNA of bacterial origin which is incorporated into the phage genome. It is thought that the rapid replication of the operons concerned leads to their outnumbering the repressor molecules available. It is possible that the same situation applies to colicins such as E1-K30 and that the induction of colicin synthesis is a direct result of induction of C-factor replication. This hypothesis would explain why after induction of colicin synthesis by thymine deprivation, the thymine must be replaced, presumably to allow DNA replication (LUZZATI and CHEVALIER, 1964). However, we must not ignore the evidence which KOHIJAMA and NOMURA (1965) obtained with a colicinogenic strain of $E.$ $coli$ K12 in which DNA synthesis is reduced by more than 97% at high temperature. On raising the temperature, colicin synthesis was induced without detectable DNA synthesis. The possible residual 3% synthesis, over one generation time, would nonetheless correspond to synthesis of about 20 C-factors per $E.$ $coli$ chromosome and it is possible that the C-factor is indeed replicating.

### Regulation of the I-P9 C-Factor

The I-P9 C-factor, the second of our two examples, differs from E1-K30 in its ability to undergo epidemic spread, giving HFC cultures, and also in its ability to act as a general promoter of genetic transfer.

MONK and CLOWES [1964 (2)] and CLOWES (1964) showed that ultraviolet irradiation of $E.$ $coli$ K12 (I-P9) leads not only to an increase in the synthesis of colicin, but also to a change from the LFC to the HFC state. Conversely they found that a standard HFC culture prepared by epidemic spread gave 150 times as many lacunae as an LFC culture when plated with a suitable strain, although the absolute numbers were low (0.003% and 0.5% of the viable count), perhaps because of the insensitivity of the lacuna assay with I-P9. Thus cells induced with ultraviolet light and cells with a newly transferred C-factor both show an increased likelihood that they can transmit genetic material and both show an increased probability of producing colicin. It is not known if the same cell can actually participate in both phenomena.

OZEKI (1965) has further clarified the nature of the HFC state (Fig. 4.2). He took advantage of the fact that if an HFC culture is mixed with a culture of LT2

(E2-P9), then after infection by I-P9 from the HFC culture, the cells of the second culture themselves become able to transfer the E2-P9 factor to a third strain, whereas before they could not. He took a culture of such cells and at intervals tested their ability to transfer the E2-P9 factor to a third strain. The results (Fig. 4.2) show that it took 20—45 min for the newly (I-P9)$^+$ strain to become a donor itself, and that for the first 100 min, as these cells multiplied, all the progeny could also act as donors but that from about 100 min the number of competent donors remained steady, while the (I-P9) (E2-P9)$^+$ cells continued to multiply. Thus, once a culture enters the LFC

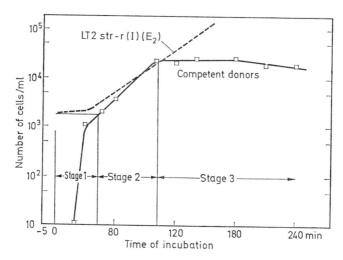

Fig. 4.2. *Phenotypic expression of the competent donor character.* An HFC (col I$^+$ col E2$^-$) donor culture was mixed with a culture of LT2 (E2-P9), at − 5 min., in the ratio 1:4. After 5 min. at 37°, the mixture was diluted (10$^{-3}$) in prewarmed fresh broth and incubated with aeration. At every 30 min., the culture was diluted twofold by prewarmed fresh broth. At intervals, the number of competent donor cells transmitting E2-P9 was measured. The broken line derives from a parallel experiment using Streptomycin-resistant LT2 (E2-P9) and is a measure of the number of the cells originally (E2-P9)$^+$, which now also carry the I-P9 C-factor. It represents the growth curve of the LT2 (E2-P9) bacteria which received I-P9 in the 5 min. before the onset of the experiment. (From OZEKI, 1965)

state, the ability to act as a donor is inherited unilineally; what is it that is inherited in this way? It is thought to be a "machinery" for the synthesis of conjugation apparatus (that it is more than a pilus with receptors was shown by destroying the receptors with periodate and demonstrating that they could be regenerated by the postulated "machinery").

This temporary derepression of promoter activity resembles that of the Lac operon on transfer from an Hfr strain to an F$^-$ strain lacking repressor (PARDEE, JACOB and MONOD, 1959). In the case of the Lac operon, it is thought that after transfer it takes some time for the synthesis of sufficient repressor molecules to prevent further function of the Lac operon. By analogy it would seem that the development of promotor activity is inhibited by a repressor encoded by a gene also on the (P9-I) epichromosome, and that after transfer, it takes time for newly synthesized repressor to shut off further colicin synthesis. Conjugation machinery already existing at this stage is simply transferred to one of the two daughter cells at each division.

Fertility derepressed (Fdr) mutants of (P9-I) have now been isolated, which confer on bacteria the ability to synthesize conjugation machinery all the time (ED-WARDS and MEYNELL, 1968; OHKI and OZEKI, 1968). These mutants presumably lack a functional gene for the repressor and the bacteria are therefore permanently in the HFC state. The existence of these mutants further supports the hypothesis that the LFC state is normally maintained by a repressor.

OHKI and OZEKI (1968) found that their Fdr mutants retained immunity to phage BF23, showing that this property is under the control of another repressor. DOWMAN and MEYNELL (1970) showed that colonies of Fdr mutants produced larger inhibition zones on a colicin sensitive indicator strain, than did the parent strain. However they attributed this to the pleiotropic effects of the mutation on cell fragility, suggesting that this led to greater release of the colicin, which is normally cell bound. Fdr mutants apparently do not produce more colicin than the parent strain and colicin production must therefore be controlled independently of fertility.

# Classification of C-Factors

In this chapter we have discussed several of the known properties of C-factors, usually using as an example E1-K30 or I-P9. However, quite a lot of information is available on other C-factors and we can now see how they compare. Much of the information is summarised in Table 3.4.

## The Specificity of the Sex-Pilus

MEYNELL and DATTA [1966 (1, 2)] and LAWN et al. (1967) compared the sex pili determined by many different epichromosomes and were able to classify most into one of two groups, those resembling the F-pilus in morphology, sensitivity to male specific phages, and also in their antigenic specificity, and those resembling the I-P9 sex pilus in these respects. It can be seen from Table 3.4 that there appears to be a correlation between pilus specificity and the colicin type determined. In each case the pilus was produced on only a small proportion of the cells; like the I-P9 pilus, their synthesis is normally repressed.

## Repressors of Promoter Activity

E1a-SL1O15 and four similar C-factors undergo epidemic spread through col-cultures like I-P9 (LEWIS and STOCKER, 1965). In these cases it is possible to see if a resident C-factor can prevent epidemic spread of another and in fact the five C-factors of type E1a, as well as E2-K317, render a cell unable to support epidemic spread of I-P9. The two E1a C-factors tested did not prevent the I-P9 C-factor being transferred to a cell and being stably inherited. They merely prevented it undergoing epidemic spread, presumably because they produce a repressor similar to that produced by I-P9 itself (LEWIS and STOCKER, 1965; MEYNELL and LAWN, 1967), and hence the transient HFC derepressed state does not occur. Likewise, Ib-P9 and Ia-CA53 prevent epidemic spread of two E1a C-factors.

It would seem that all 5 E1a C-factors, I-P9, E2-K317 and Ia-CA53 produce a similar repressor to control promotor activity. It will be noted that they all determine I-type pili with the exception of E2-K317, which is not known to confer any fertility or to determine a pilus.

CLOWES, HAUSSMAN, NISIOKA and MITANI (1969) found four C-factors for colicins of type B to be capable of epidemic spread (Table 3.4). All four determine F-type sex pili. All are able to coexist in a cell with the K12 sex-factor and three of them repress its promotor activity, presumably by repressing sex pilus synthesis (CLOWES et al., 1969; PUIG and NAGEL DE ZWAIG, 1964). Some R-factors, known as fi⁺ (fertility inhibition) are also able to repress the K12 sex-factor in this way (WATANABE and FUKASAWA, 1962) and all presumably synthesize a similar repressor. In the case of B-CA98 and B-K77, fertility derepressed mutants have been obtained similar to those obtained from I-P9 in that the majority of the cells carry the sex pilus and transfer the epichromosome. The existence of these mutants, as for I-P9, confirms the existence of a repressor controlling promotor activity. The mutants, as one might expect, are unable to repress the promotor activity of the K12 sex-factor.

### The Rec Gene and Induction of Various C-Factors

We saw that *E. coli* K12 (E1-K30) did not produce colicin if the strain was also RecA⁻ [HERSCHMAN and HELINSKI, 1967 (1)]. The same authors also showed that this applies to E2-P9 in RecA⁻ bacteria but that *E. coli* K12 (VI-K94) could produce colicin in the presence of the RecA⁻ allele. It is of interest that C-factors affected by the RecA gene are inducible, whereas VI-K94 is variously reported as not inducible [HERSCHMAN and HELINSKI, 1967 (1)] or barely inducible (KAHN, 1968).

### Incompatibility between Epichromosomes and Entry Exclusion

In some cases it is possible to transfer, one after the other, two or more epichromosomes to a given bacterium and to prepare stable strains carrying more than one epichromosome. We have already referred to some of the interactions observed in such strains. However, different F' factors, each derived from the K12 sex-factor, cannot normally coexist in the same cell (ECHOLS, 1963; DE HANN and STOUTHAMER, 1963). Likewise two distinguishable derivatives of the VI-K94 C-factor could not coexist in the same cell (NAGEL DE ZWAIG, 1966). NOVICK (1969) has discussed this phenomenon, which may be due to competition between two epichromosomes for a particular site within the cell. It is therefore of interest that VI-K94 and the *E. coli* K12 sex-factor are mutually incompatible and hence perhaps compete for the one type of site (MACFARREN and CLOWES, 1967; NAGEL DE ZWAIG and ANTON, 1964, 1965). VI-K94 and the K12 sex-factor can each be efficiently transmitted to cells already carrying the other but are not replicated (MACFARREN and CLOWES, 1967). However, some but not all Hfr strains are able to carry the VI-K94 C-factor.

The K12 sex-factor and V-K30 are also incompatible in *E. coli* K12, but in this case the incompatibility remains even if the sex-factor is incorporated to give an Hfr strain. An incoming V-K30 epichromosome can often displace a resident K12 sex-factor (MacFarren and Clowes, 1967). Kato and Hanaoka (1962) observed a similar type of interaction between certain R-factors and C-factors of *E. coli* K235; when the R-factors were introduced into *E. coli* K12 (K-K235) (X-K235), a proportion of the progeny were found to have lost either their ability to make colicin X or colicin K, presumably due to elimination of the K-K235 C-factor or the presumed X-K235 C-factor.

Mutual incompatibility suggests that the epichromosomes involved have some common requirement in the cell and indicates relatedness.

In addition to incompatibility, some epichromosomes can exclude others from entering. Thus a resident K12 sex-factor will exclude an incoming K12 sex-factor (Echols, 1963). This phenomenon has been demonstrated for Ib-P9 (Meynell, 1969). These various interactions between epichromosomes have been reviewed by Novick (1969).

For technical reasons, it is often difficult to distinguish between incompatibility and exclusion. The "superinfection immunity" often observed between C-factors probably includes examples of both phenomena (Meynell, 1969; Meynell, Meynell and Datta, 1968; Novick, 1969).

### Immunity to Bacteriocins

Strains which produce bacteriocins are, in general, resistant to those bacteriocins, and this applies also to strains which have been made colicinogenic after receiving a C-factor from another strain [Fredericq, 1956 (2)]. *E. coli* K12, sensitive to all colicins, becomes immune to a colicin for which it is colicinogenic; *E. coli* K12 (E1-K30) is thus immune to colicin E1-K30, and also, in fact, to many other colicins of type E, which, as we saw in Chapter 1, are defined on this basis as being of sub-type E1. The specificity of this immunity does not necessarily correspond to that determined by resistant mutants and used as the major basis for classification of colicins. The colicins of type E, as we have seen, are divided on the basis of immunity into types E1, E2, E3 and E4. The specificity of immunity of colicin I has also been studied and makes it possible to recognize sub-types Ia and Ib, and immunity to colicin V shows a similar pattern, as we saw in Chapter 1. Other than in these examples the specificity of immunity has not been studied in detail. Immunity of this type is not always absolute and colicinogenic strains may be sensitive to a high concentration of their own colicin. C-factors of type Ib and also E2-K317 are of interest as these not only confer immunity to the corresponding colicins (Ib *or* E2), but cause infections by phages BF23 or T5 to abort (Stocker, 1966; Strobel and Nomura, 1966; Taizo and Ozeki, 1968).

The immunity conferred by a C-factor is often compared to that conferred by a temperate phage against superinfection by other phages of the same type. However, it is worth noting that there are significant differences. The temperate bacteriophage λ, on adsorption to a sensitive cell, depresses host RNA and protein synthesis (Terzi and Levinthal, 1967). The immunity conferred by λ prophage is to replication of superinfecting λ phage DNA and does not prevent the depression of host RNA and protein synthesis referred to above. The immunity conferred by a C-factor, on the other hand, is against the metabolic effects which otherwise result from the adsorption of a colicin molecule. However, immunity to some defective bacterio-phages resembles that to colicins (SUBBAIAH et al., 1965). These defective phages have no DNA and cause lethal metabolic disturbances of the sensitive cell. Lysogenic strains are immune to these effects.

Immunity to colicin E3-CA38 is due to an intracellular inhibitor of colicin action (BOWMAN, SIDIKARO and NOMURA, 1971), which is discussed in Chap. 6.

Immunity would seem to be an essential property to be conferred by a C-factor, as otherwise the colicinogenic cell would be expected to die, but E. coli 15, on induction, produces a defective phage for which the only known sensitive strain is E. coli 15 itself (RYAN et al., 1955) and a similar situation applies for the colicin described by COCITO and VANDERMUELEN-COCITO (1958). We shall see in the Appendix that whereas bacteriocins of some groups, such as enterococcins and pyo-cins, do not in general act on the producing strain, this does not apply to others such as megacin A or Pesticin II. In this chapter, we need only note this immunity as a ge-netic property conferred by C-factors which may or may not be related to the other types of repression we have discussed; in Chapter 6 when we consider the mode of action of bacteriocins, we will find this immunity of considerable relevance, although it is not well understood.

## Stability and Elimination of Epichromosomes

C-factors are generally very stable properties of a clone, but may occasionally be lost [HAUDUROY and PAPAVASSILIOU, 1962 (2); MONDOLFO and CEPPELLINI, 1950; PAPAVASSILIOU, 1962].

It was first shown for the E. coli K12 sex-factor that an epichromosome may be more susceptible to certain chemicals than the main chromosome, and E. coli K12 F+ strains, grown in the presence of certain metals or acridine orange, may become F− (HIROTA, 1960). This observation has been extended to C-factors. The bacteria are said to have been "cured" of the sex-factor or C-factor which seems to have been completely eliminated by this treatment. Colicinogenic strains which have been cured by acridine orange or other means include K12 (V-K30) (MACFARREN and CLOWES, 1967), K12 (V-K94) (KAHN and HELINSKI, 1964; MACFARREN and CLOWES, 1967) and CA31 (A-CA31) (LORKEIWICZ et al., 1965). LORKEIWICZ found that nitrous acid would render E. coli CA53 or E. coli CA31 non-colicinogenic, whereas by acridine orange he could cure CA31 but not CA53. CA53 is naturally colicinogenic for colicin I, and CA31 for colicin A. However, not all C-factors can be removed by these means and several C-factors have resisted attempts to eliminate them, e.g. K12 (E1-K30) (KAHN and HELINSKI, 1964), Providence (D-CA23) and Providence (J-CA62) (COETZEE, 1964) and K12 (B-K260) (MAYNE, 1965). CLOWES (1964) made the interesting observation that I-P9 and E1-K30, although not eliminated from E. coli K12 by acridine orange, could be eliminated from a thymine-requiring strain by thymine deprivation, and MACFARREN and CLOWES (1967) found that VI-K94 and V-K30 were also eliminated. SMIT et al. (1968) reported that Proteus morgani MR336 could be cured by acridine orange of its ability to produce a morganocin.

It is not known how these various agents or conditions cause some epichromosomes to be eliminated, but they perhaps inhibit replication of some epichromosomes more effectively than replication of the main chromosome.

## Conclusions

This chapter has been largely concerned with properties such as regulation of epichromosomal DNA replication, and interactions between epichromosomes, which have been studied for many epichromosomes other than C-factors. We have confined our discussion almost entirely to C-factors and find that they vary amongst themselves in these properties. The properties of other epichromosomes have been remarked upon in passing but are treated in detail elsewhere (NOVICK, 1969; and also several of the papers in Ciba Foundation Symposium, 1969).

The genetic properties of C-factors which relate more directly to colicin production are not well understood. However we have seen that induction of colicin synthesis may be a corollary of the location of the structural gene for a colicin on an epichromosome. C-factors also confer immunity to colicins related to that encoded, but the nature of this immunity is understood only for colicin E3-CA38 where it is due to an intracellular inhibitor.

# Chapter 5

# Mode of Action — The Adsorption of Bacteriocins

## How Bacteriocins Kill

The interest of bacteriocins to molecular biology centres largely on their remarkable mode of action. As yet, too few have been studied for us to be able to generalize with much confidence but, as we shall see, the evidence adduced in these studies certainly suggests that one molecule, after adsorption to a specific receptor on the bacterial surface, then kills that bacterium. Different bacteriocins inhibit DNA synthesis or protein synthesis, and some inhibit a wide range of activities, such as DNA, RNA and protein synthesis, together with permease function and are thought to have their primary effect on the deployment of energy by the bacterium. This is clearly different to the action of the better-known typical antibiotics, substances usually of relatively low molecular weight which enter the cell and either inhibit enzyme function by acting as analogues of the substrate or, in other cases, bind to DNA, blocking its template function.

The action of bacteriocins is reminiscent of some of the bacterial protein exotoxins, such as that of *Clostridium botulinum* (VAN HEYNINGEN, 1950), which adsorb strongly to the sensitive mammalian cell, presumably to some sort of receptor, and are very efficient at destroying specific cell functions; in no such case is the mode of action understood at the molecular level.

## The Colicin Receptor

FREDERICQ [1946 (4)] showed that if colicin K was added to a suspension of sensitive cells, only about 0.2% of these could give rise to colonies if plated out after 2 min. Because of this very rapid rate of killing, he came to the conclusion, still generally held, that killing involves binding of colicin to cells and that the final killing process may take much longer. The most likely site for this irreversible binding is the cell surface, and support for this hypothesis came early from the observation that antibacterial sera can protect bacteria from colicins (BORDET, 1948; BORDET and BEUMER, 1948; MAYR-HARTING, 1964) even if the bacteria used to elicit the antisera did not carry the colicin receptor. The antisera presumably bind to the bacterial surface and are thought to prevent colicin adsorption by steric hindrance, making the receptors not available.

Antisera similarly prevent bacteriophage adsorption but not the action of antibiotics such as penicillin. Further support for surface adsorption comes from work on "trypsin rescue". Using suitably low concentrations of colicin, it is possible to rescue with trypsin or chymotrypsin cells which have already adsorbed colicin and are doomed to die (HULL and REEVES, unpublished data; NOMURA and MAEDA, 1965; NOMURA and NAKAMURA, 1962; REYNOLDS and REEVES, 1963). In the case of colicin K-K235, rescue of a proportion of the cells by trypsin can be effected long after adsorption (NOMURA and NAKAMURA, 1962). Colicins E2-CA42 (REYNOLDS and REEVES, 1963), E3-CA38 (NOMURA and MAEDA, 1965), and pesticin 1 (ELGAT and BEN-GURION, 1969) must normally be removed with trypsin soon after adsorption for effective rescue. However, in the presence of 2:4 dinitrophenol or cyanide, colicins of type E2 produce no detectable effect on DNA (HULL and REEVES, unpublished data; NOMURA and MAEDA, 1965), and trypsin or chymotrypsin rescue is effective for an hour or more (HULL and REEVES, unpublished data; REYNOLDS and REEVES, 1963). The ability of proteolytic enzymes to rescue cells from the effects of adsorbed bacteriocins tells us that the receptor is on the surface and exposed to trypsin.

Of importance here is the work of MAEDA and NOMURA (1966), who, using radioactive colicin E2-P9, disrupted and fractionated cells after 60 min contact with colicin and found that only 1% of the colicin was soluble. Eighty-one percent was associated with the cell envelope, while 8% was with the ribosome fraction (100,000 × g precipitate), and 10% with unbroken cells and large fragments. It is clear that very little of the colicin becomes soluble during killing although as we shall see later it may in fact be the small amount which does become soluble which produces the lethal effects.

# The Quantitation and Specificity of Colicin Adsorption

## Specific Adsorption of Colicin by Sensitive Bacteria

The adsorption of colicin by sensitive bacteria can be quantitated by mixing bacterial cells and bacteriocin for a while, removing the bacteria by centrifugation and assaying the residual colicin. BORDET and BEUMER (1948, 1949, 1951) found that extracts of E. coli φ (or CA81) sensitive to colicin V-CA7 were able to neutralise the colicin action, whereas similar extracts of either an insensitive unrelated strain, or resistant mutants of the sensitive strain, did not neutralise the activity. HAMON and PERON (1960) studied several colicin and pyocin preparations and reported that with colicin suitably diluted, sensitive bacteria could neutralise the activity, whereas colicin resistant mutants could not.

More detailed studies were carried out on the adsorption of two colicins of type E2. MAYR-HARTING (1964) and MAYR-HARTING and SHIMELD (1965) studied the adsorption of E2-P9 to E. coli C6 and S. sonnei, and REEVES [1965 (1)] the adsorption of E2-CA42 to E. coli K12. Both found over a suitable range of bacterial and colicin concentrations that the amount of colicin adsorbed was proportional to the number of bacteria in the adsorbing suspension. Furthermore, resistant mutants did not adsorb colicin at all (Fig. 5.1), showing that the adsorption of colicin by sensitive cells is specific and associated with colicin action. MAEDA and NOMURA (1966) used radioactive colicin to study adsorption and confirmed that colicin E2-P9 adsorbed to sensitive cells but not to a resistant mutant. With this technique it was possible to show that adsorption of nonradioactive colicin E3-CA38 or E2-P9 would prevent subsequent adsorption of a radioactive preparation of the other colicin, confirming the opinion discussed in Chapter 1 that the two colicins adsorb on the same receptor. NOMURA was also able to show that radioactive colicin, once adsorbed, was not displaced by excess "cold" colicin during various extraction procedures, once again confirming the belief that the colicin was bound irreversibly. The resistant mutants we have referred to are those discussed in Chapter 1, which

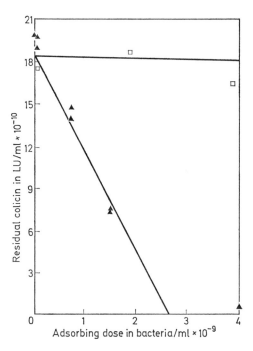

Fig. 5.1. *Adsorption of colicin E2-CA42 by E. coli K12.* Log phase bacteria of a substrain of *E. coli* K12 (▲ ——— ▲) and a mutant resistant to the colicin (□ ——— □) suspended in a salt solution were diluted to the concentrations indicated. To each dilution was added an equal amount of colicin in minimal salts. After 1 h incubation at 37 °C the bacteria were removed by centrifugation and residual colicin assayed. From the slope of the first it can be seen that each bacterium adsorbed about 70 LU of colicin and in a series of such experiments an average value of 60 LU/bacterium was obtained (After REEVES, 196 S)

define colicin type E. They affect colicin adsorption and are usually taken to lack a receptor common to all colicins of type E. However, BHATTACHARYYA, WENDT, WHITNEY and SILVER (1970) suggest that these mutations merely restrict access of colicin to an unaltered receptor, while HAMON and PERON [1966 (1)] consider that there are at least two different types of receptor for colicin E.

Apart from this work on colicins, there has been very little work on adsorption of bacteriocins. We have already mentioned HAMON's observations on pyocins and should mention HOLLAND's data on megacin A-216. HOLLAND (1962) showed that *B. megaterium* 207 adsorbed megacin A-216 from solution, and that other bacteria such as *E. coli* or *B. megaterium* strain 216 (the producing strain which is, however, sensitive to its own megacin) adsorbed less megacin activity. It is difficult to know if the adsorption of a small amount of protein (probably less than 0.1 % of the bacterial weight) is specifically to receptors. Unfortunately, it has not been possible to isolate megacin-resistant mutants and the use of mutants lacking receptors would provide the only good estimate of any non-specific adsorption to the cell surface. However, now that megacin A-216 has been shown to be an enzyme (Ozaki *et al.*, 1966) (see Chap. 3), it is doubtful if one can really talk in terms of "receptor" and specific "adsorption" in this case. Kageyama, Ikeda and Egami (1964) studied the adsorption of the "pyocin" produced by *P. aeruginosa* R using radioactive $^{35}$S labelled pyocin and found that it was adsorbed to sensitive but not insensitive strains. However, as we saw in Chap. 2, their "pyocin" consists, in our definition, of defective bacteriophages.

## The Quantitation of Adsorption

If bacteriocins adsorb to specific receptors on the bacterial cell, then it is reasonable to ask how many of these receptors exist and how many colicin molecules must be adsorbed in order to kill a sensitive cell. The number of molecules required to kill a cell can be estimated by studying the kinetics of adsorption measured as killing.

The first study of this aspect of the action of bacteriocins was included in the pioneering paper by Jacob *et al.* (1952), and since then there have been several papers concerned largely or partly with the kinetics of adsorption. The method used in all such studies to date is to mix various concentrations of colicin and bacteria and incubate them together. At intervals samples are taken, diluted to prevent further appreciable adsorption of colicin, and viable counts done. In this way one determines the rate at which bacteria accumulate sufficient adsorbed bacteriocin to kill them. Recently Shannon and Hedges (1967) have improved the sampling techniques used and can determine the viable count at 10 sec intervals. In Fig. 5.2 we see the results of one of their experiments. The viable count falls from the time colicin is added at a rate dependent on the concentration of the colicin. The initial rate of killing soon decreases, but if we look only at the initial slope, then we find a linear dependence on colicin concentration and in Fig. 5.3 we see such a plot for a different set of data using colicin E2-CA42. How are we to interpret this? Let us confine our attention for a moment to the generally accepted hypothesis and see how the data fit this. If a single molecule of colicin can kill a cell, then the rate of killing will be proportional to the rate at which colicin adsorbs to cells, and this will follow ordinary reaction kinetics.

$$- \frac{dB}{dt} = kBC$$

where B is the concentration of viable bacteria and C the concentration of colicin. If the log proportion of bacteria surviving (log $B_t/B_o$) is plotted against time we would expect a straight line with a slope proportional to C. This occurs for the initial slopes in Fig. 5.2. We have in fact observed what is called "single-hit" kinetics.

We should note here that we have not had to specify either the probability of an adsorbed colicin molecule killing a cell which could be any value up to 1, or the number of receptors per cell, or even, while we confine our attention to the initial slope, that the cell population is homogeneous in these respects.

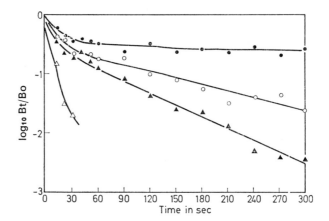

Fig. 5.2. *Kinetics of colicin killing.* Colicin E2-P9 at various dilutions was added to a suspension of a sensitive strain of *E. coli* at 0° and at intervals a sample taken, diluted and the number of viable bacteria determined by plate count and expressed as the Log$_{10}$ of the proportion surviving. Colicin diluted ●———● 1:16; ○———○ 1:8; ▲———▲ 1:4; △———△ 1:2. [From SHANNON and HEDGES (1967)]

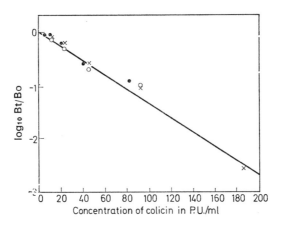

Fig. 5.3. *Dependence of initial killing rate on colicin concentration.* Colicin E2-CA42 at various concentrations was added to a suspension of *E. coli* K12 and after one minute the viable count determined. This was taken as a measure of the initial rate of killing. Three separate experiments are shown. [From REEVES, 1965 (1)]

## The Single-Hit Hypothesis

We have shown by plotting log Bt/Bo against t, and looking at the dependence of the initial slope on colicin concentration, that we can recognize a "single-hit" curve. The slope of the curve at any particular point will be dependent on (1) the number of receptors per bacterium S, (2) the concentration of colicin C and (3) the reaction constant for the reaction between the colicin and receptor. If all bacteria have the same number of receptors (S is constant) then the only variable will be C, and if we use an excess of colicin such that throughout the experiment the amount of colicin adsorbed has only a negligible effect on the amount of colicin present, then we would expect the slope to remain constant throughout. However, if we look at Fig. 5.4, in which such an excess of colicin was present, we see that the slope does, in fact, change as the killing progresses, and this is thought to be due to heterogeneity in the number of receptors per cell. The initial slope is determined by the average number of

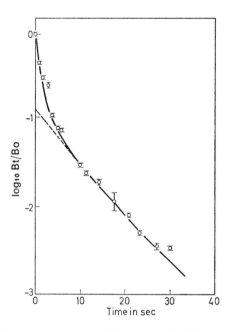

Fig. 5.4. *Analysis of a typical survivor plot.* The data obtained in an experiment similar to that described in Fig. 5.2 with the 95% confidence limits indicated. The line is that predicted for a hypothetical population having the following distributions of specific lethal receptors per cell: 2, 14%; 10, 30%; 12, 50%; 15, 6%. (From SHANNON and HEDGES, 1967). For further analysis see text

receptors per bacterium, and the final slope is due to the last surviving bacteria; i.e., those with the least number of receptors (HEDGES, 1966; SHANNON and HEDGES, 1967). The two slopes differ by a factor of 6, suggesting a sixfold difference between the average and the least number of receptors per cell in the population. The change of slope of killing curves has been commented on before, e.g. by HAMON and PERON (1960).

It should be noted that the heterogeneity indicated by the change of slope is in the number of receptors able to take part in effective adsorption, that is adsorption which proceeds to a lethal event. Any receptors to which colicin can adsorb, but which do not allow colicin action to proceed further will not be scored in this assay.

## The Multi-Hit Hypothesis

The logical alternative to the "single-hit" hypothesis for colicin killing is, of course, the "multi-hit" hypothesis (i.e., that killing results from the cumulative action of more than one colicin molecule). The actual kinetics expected from a multi-hit model will depend both on the number of "hits" required to kill, and on the number of receptors on each cell, but in all cases one would expect a shoulder on the plot of log Bt/Bo against time. No shoulder is usually reported but the data are rarely critically examined to eliminate the multi-hit hypothesis. However, this can be done and, for example, the data plotted in Fig. 5.3 do not fit the curves expected on the hypothesis that 2 or 3 molecules are required to kill [REEVES, 1965 (1)]. In this case at least it seems that the single-hit hypothesis is the only reasonable one, and few authors have ever obtained data to support a multi-hit hypothesis.

## The Lethal Unit

Where there is evidence that cell death can result from the adsorption of a single molecule of colicin, then it is reasonable and usual to refer to the average amount of colicin required to kill a single cell as a lethal unit (L. U.) or a killing unit. We should note that the lethal unit corresponds to a molecule only if each molecule which adsorbs has the maximum probability of 1 of killing a cell. The number of molecules which constitute a lethal unit will obviously depend on the test conditions if these affect the probability of an adsorbed molecule killing a cell. Nonetheless, the amount of bacteriocin used in an experiment is often expressed in lethal units, and this is readily determined by taking the proportion of survivors after colicin treatment as being the proportion of cells which have not adsorbed a lethal unit and, using the Poisson distribution, calculating the average number of lethal units per bacterium and hence the titre of a colicin solution in lethal units. Under standard conditions the assay is reproducible.

## The Amount of Colicin Adsorbed per Cell at Saturation

If the amount of colicin adsorbed per sensitive cell is measured in lethal units, then one has an estimate of the number of receptors per bacterial cell. In all such experiments an important control is the use of a receptorless resistant mutant which does not adsorb colicin in any specific way and gives an estimate of any non-specific adsorption.

MAYR-HARTING (1964), working with *E. coli* C6 and colicin E2-P9, found that each cell could adsorb 11 L. U. of colicin. REEVES [1965 (1)] measured the amount of E2-CA42 adsorbed by *E. coli* K12. We have already looked at these data (Fig. 5.1), which show that *E. coli* K12 adsorbed about 30—90 L. U. per cell. E2-CA42 is one of the colicins which have been purified (see Chap. 2), and it was possible to estimate that in the most active preparation obtained there was about $10^5$ Dalton associated with each lethal unit, a figure which agrees well with the estimated molecular weight of $6 \times 10^4$. This suggests that there are 30—90 receptors per cell.

MAEDA and NOMURA (1966) found that under their conditions and using colicins E2-P9 and E3-CA38, *E. coli* K12 could adsorb 20—30 L. U. per cell. They were using radioactive colicin and estimated that in this case there were from 40—300 molecules adsorbed per lethal unit, depending on the experimental conditions and preparations used. Since all the radioactivity seemed to be specifically adsorbed, they concluded that there are 2—3,000 colicin E receptors per cell. This is much higher than the 30 to 90 receptors per cell observed by REEVES [1965 (1)], also using a colicin of type E2. The reason for this large difference is not known.

In a comparative study CAVARD and BARBU (1966, 1970) demonstrated a considerable variation in the number of L. U. of different colicins which could be adsorbed per cell. Each cell could adsorb 600 L. U. of K-K235, 60 L. U. of E2-K317 and 1—3 L. U. of A-CA31.

We have already seen that the analysis of the killing kinetics suggests that there may be considerable variation in the number of receptors on the cells of a given population. MAYR-HARTING and SHIMELD (1965) found that the average number of receptors per cell can also be markedly affected by the growth conditions. *E. coli* C6 grown on ox heart infusion agar and washed off could adsorb an average of 11 L. U. per cell of E2-P9, whereas those grown in the same nutrient medium in liquid form adsorbed less than 10% as much. The broth-grown bacteria were also less sensitive to the colicin. The effect was specific in that liquid media other than ox heart infusion broth gave fully sensitive bacteria.

## Number of Molecules per Lethal Unit

The data of MAEDA and NOMURA (1966) which we discussed above showed clearly that a lethal unit may comprise much more than one molecule. Also DE GRAAF *et al.* (1971) showed that a lethal unit of cloacin DF13 consists of about 12 molecules although single-hit killing kinetics apply (DE GRAAF, SPANJAERDT-SPECKMAN and STOUTHAMER, 1969). There are several possible explanations for this which are compatible with the single-hit killing kinetics discussed earlier. It may be that each molecule adsorbed has an equal but low probability of killing the cell, or that some receptors are not able to take part in initiating the lethal event. It is also possible that colicin preparations include a proportion of inactive molecules which can nonetheless adsorb. Whatever the true explanation, it is likely that the number of molecules which comprise a lethal unit will vary either with the colicin preparation or with the growth conditions, etc., of the sensitive culture. There is already evidence that in some cases culture conditions can affect the probability that an adsorbed colicin molecule will kill a cell. NOMURA, in the published discussion of the paper by MAYR-HARTING and SHIMELD (1965), says that the concentration of $Mg^{++}$ ions can have an effect on the lethality of pre-adsorbed colicin. $5 - 10 \times 10^{-3}$ M was optimal for killing. Likewise, BEPPU and ARIMA (1967) found that if cells which had adsorbed E2-P9 or K-K235 were subsequently incubated in the presence of 0.8 M NaCl or sucrose, they were significantly protected from the lethal effects of these colicins and also from the DNA degradation caused by E2-P9.

The response is also under genetic control. NAGEL DE ZWAIG and LURIA (1967) find that tolerant mutants (see Chap. 6), while adsorbing colicin normally and exhibiting single-hit killing kinetics, are only killed by an adsorbed colicin molecule with $5 - 10\%$ the efficiency with which wild type cells are killed.

## The Reaction between Colicin and Receptor

The adsorption of bacteriophages to their specific receptors has been shown in several studies to depend on the ionic environment and in some cases to require specific adsorption co-factors, such as tryptophan (see STENT, 1963 for review). Similar studies on the adsorption of colicin E2-CA42 (REYNOLDS and REEVES, 1969) have

Table 5.1. *Adsorption coefficient of colicin E 2-CA42 onto E. coli K12 in various media*
[After REYNOLDS (1966)]

| Medium | Adsorption constant ($K \times 10^{-11}$) Expt. | | | |
| --- | --- | --- | --- | --- |
| | 1 | 2 | 3 | 4 |
| Nutrient broth | 5.9 | 6.8 | | |
| Minimal salts solution | | 24.4 | | |
| Tris buffer pH 7 (0.05 M) | 8.9 | | 8.9 | |
| Tris buffer pH 7 (0.05 M) $+10^{-4}$ M Mg$^{++}$ | | | 20.0 | |
| Tris buffer pH 7 (0.05 M) $+10^{-3}$ M Mg$^{++}$ | | | 23.5 | |
| Tris buffer pH 7 (0.05 M) $+10^{-2}$ M Mg$^{++}$ | | | 9.8 | |
| Phosphate buffer pH 7.0 (0.1 M K$^{+}$) | | | | 24.4 |
| Phosphate buffer pH 7.0 (0.1 M K$^{+}$) $+10^{-4}$M Zn$^{++}$ | | | | 8.9 |
| Phosphate buffer pH 7.0 (0.1 M K$^{+}$) $+5 \times 10^{-4}$M Zn$^{++}$ | | | | 1.5 |
| Phosphate buffer pH 7.0 (0.1 M K$^{+}$) $+10^{-3}$M Zn$^{++}$ | | | | 0.75 |

Log phase cells of a strain of *E. coli* K 12 grown in nutrient broth were washed in deionized water and diluted ten fold into the requisite media. The proportion of cells surviving at suitable time intervals was measured and the survival at 1 min. obtained by extrapolation and used to determine the adsorption coefficient.

shown a similar dependence on the ionic environment. The initial rate of killing can be used to calculate the adsorption constant:

$$K = \frac{1}{Ct} \cdot \ln \frac{Bo}{Bt}$$

which in these experiments was based on the number of bacteria killed in the first minute after colicin addition. $C$, the concentration of colicin was measured in lethal units. As can be seen from Table 5.1, a Mg$^{++}$ concentration of $10^{-3}$ M is optimal, whereas Zn$^{++}$ is inhibitory to adsorption. These results are comparable to those obtained for bacteriophage adsorption (GAREN and PUCK, 1951). It was also found that adsorption was optimal at pH 7—8.

## The Role of the Colicin Receptor

Colicin, after adsorption to a surface receptor, can proceed to have a lethal effect. Until recently it had been generally assumed that colicins exerted this lethal effect while still adsorbed to the receptor. We have seen that in the case of colicin K-K235, trypsin rescue experiments support this hypothesis. However, the recent observation that colicin E3-CA38 can act on ribosomes *in vitro* [Boon, 1971; Bowman, *et al.*, 1971 (2)] suggests that it acts from within the cell (see Chap. 6 for discussion). If this is indeed so the adsorption to receptor must be followed by transport of the colicin molecule across the cell envelope, if the colicin is to have any effect.

If colicins of type E2, which have been used for many of the experiments described in this chapter, also act from within the cell, then we can draw some further conclusions on the colicin-receptor interaction. The large amount of colicin which for some reason remains on the surface of the cell could not participate in the lethal effect. This would be the explanation for one L. U. comprising many colicin molecules. We could perhaps also expect that in the presence of cyanide, etc., colicin could not enter the cell, thus accounting for the trypsin rescue experiments with colicins of type E.

We have seen that there is considerable evidence that colicin adsorbs irreversibly to cells, and that a surface receptor is involved. Likewise there is a maximum amount of colicin which can be adsorbed per cell. However, none of this incompatible with the hypothesis that colicin enters the cell, as each receptor may be able to transfer only a limited amount of colicin.

# The Nature of the Receptor

## Cross-Resistance between Colicins and Bacteriophages

Colicins and bacteriophages have similar types of activity spectra and are both inducible. These and other similarities may have provided encouragement to look for further analogies, and in 1949 Fredericq and Gratia demonstrated that particles of the two types could adsorb to common receptors. They used 8 mutants of *S. sonnei* E90 made successively resistant to colicins V, E, J, K, B, D, S2 and S1; that is, their first mutant resisted colicin V, the second was derived from it and resisted colicin V and E, and so on. When 58 different bacteriophages active on E90 were tested against the 8 mutants they found two groups of phages which are of particular interest to us, 17 in what they called group II which did not act on E90 after mutation to colicin E resistance, and 6 in what they called group III which did not act on E90 after mutation to colicin K resistance. The other 35 phages in general acted on all the mutants. It seemed then that sensitivity to colicin E and phages of group II could be lost by a single-step mutation, and likewise for colicin K and phages of group III. In later papers [Fredericq, 1949 (1, 2); Fredericq and Gratia, 1950] it was shown that most mutants selected by colicin E or phages of group II were resistant to both and likewise for colicins of type K and phages of group III.

Other examples of cross-resistance between colicins and bacteriophages involve colicins B (*not* the typical colicin B) and M, which give cross-resistance with phage T3 or with phages T1, T5, and T7 respectively [Cocito and Vandermuelen-Cocito, 1958; Fredericq, 1950 (4), 1951 (1, 2, 4, 5); Fredericq and Smarda, 1970]. In the latter example there is also cross-resistance with colicin C. Although only a few examples of cross-resistance between colicins and bacteriophages have been reported, this cannot be considered a good estimate of the number of examples which exist, as there have been no systematic searches for cross-resistance since FREDERICQ found the first two examples, and in his initial study he used only 8 colicins and one resistant mutant for each. Most of the subsequent work has been on the cases discovered at that time.

## The Bacterial Cell Wall and Its Receptors

Most bacteria are enclosed within a rigid cell wall and this may well be the location of the receptors which interest us. We shall very briefly discuss its structure, but fuller accounts are given by GHUYSEN, STROMINGER and TIPPER (1968), LUDERITZ, JANN and WHEAT (1968) and SALTON (1964). In the gram-positive bacteria the major wall component is a rigid cross-linked mucopeptide, and other components, such as teichoic acids, teichuronic acids, polysaccharides or proteins, are probably attached to it.

The cell walls of gram-negative bacteria are more complex, and although some of the components are well characterised in chemical terms, the overall structure of the wall is not yet fully elucidated. DE PETRIS (1967) discusses much of the earlier data and presents good electronmicrographs showing that outside the cytoplasmic membrane there is first the rigid mucopeptide layer, and outside this a triple-layered structure resembling a typical unit membrane. This outer unit membrane apparently contains the lipopolysaccharide of the O antigen and a considerable amount of protein.

The outer parts of the gram-negative cell wall have sometimes been thought to consist of two discrete layers of lipopolysaccharide and lipoprotein (BAYER and ANDERSON, 1965; MARTIN, 1963; MURRAY, STEED and ELSON, 1965), but as DE PETRIS (1967) points out, the evidence for such discrete layers is not conclusive, and the arrangement of the lipid, polysaccharide and proteins of the cell wall outside the mucopeptide layer is still unresolved.

## Bacteriophage Receptors

Whereas little work has been done on bacteriocin receptors, there have been many attempts to study those of bacteriophages and much has been learnt, although probably only in a few cases is the chemical basis of the receptor specificity properly known. Some bacteriophages are known to adsorb to flagellae (MEYNELL, 1961) or to sex pili (BRINTON, 1965; MEYNELL and LAWN, 1967) but most probably adsorb to receptors on the cell wall. Amongst the gram-positive bacteria there is evidence that either teichoic acids or mucopeptide, or both, may be involved in neutralising particular phages (CHATTERJEE, 1969; CHATTERJEE, MIRELMAN, SINGER and PARK, 1969; MURAYAMA, KATONI and KATO, 1968; ROSATO and CAMERON, 1964; YOUNG, 1967), while VIDAVER and BROCK (1966) found a protein-carbohydrate complex to be involved. The ability of a material to neutralise a phage presumably indicates that it contains the receptor for that phage.

Amongst gram-negative bacteria the involvement of cell wall components as receptors for phages T3, T4 and T7 has been shown both by neutralisation of phages by particular cell wall fractions (JESAITIS and GOEBEL, 1953; GOEBEL and JESAITIS, 1952, 1953), and by studying phage-resistant mutants of *E. coli* (WEIDEL, KOCH and LOHSS, 1954) or *S. sonnei* (GOEBEL and JESAITIS, 1952). All the evidence suggested that the lipopolysaccharide of the Boivin antigen contains the receptors for phages T3, T4 and T7, but the precise chemical nature of the receptor specifity is not yet known.

More recently several receptors have been shown to be on the O antigen. Several *Salmonella* phages apparently adsorb to specific parts of the O antigen (LUDERITZ et al., 1968; WILLKINSON and STOCKER, 1968), and in the case of the Felix 0-1 phage it seems that the receptor specificity is determined by an N-acetyglucosamine in the core-fraction (LINDBERG and HOLME, 1969).

The receptor specificity for other phages resides in the specific side chains of the O antigen and those for phages e[15] and e[34] are particularly well documented. The receptor activity for these phages correlates absolutely with the presence of O antigens 10 and 15 respectively, which in turn are determined by the presence of $\alpha$ acetyl galactose or $\beta$ galactose respectively, as terminal sugars in the repeating unit of the antigen (UETAKE, LURIA and BURROWS, 1958; LUDERITZ *et al.*, 1966). In this case the molecular nature of the receptor specificity seems well defined, and has in fact been directly demonstrated by *in vitro* adsorption studies (LOSICK and ROBBINS, 1969). LE MINOR (1969) described two other phages, $45_2$ and $45_4$, which can lyse only cells carrying the O antigen $45_{33}$, which perhaps constitutes the receptor.

In *E. coli* K12 the O antigen carries the receptor for phage C21 and again the specificity resides in a particular part of the antigen (KALCKAR, LAURSEN and RAPIN, 1966; RAPIN, KALCKAR and ALBERICO, 1968).

The receptor for the Vi phages of *Salmonella typhi* is the Vi antigen, which consists of a polymer of N-acetyl D-galactosamine uronic acid, which is partly O-acetylated (see LUDERITZ *et al.*, 1968, for a review). The Vi phage can be adsorbed by the purified antigen (BARON, FORMAL and SPILMAN, 1955), but the majority of the phages adsorb reversibly and can be released as infective units (TAYLOR and TAYLOR, 1963). O-acetyl groups are essential for adsorption and they and some N-acetyl groups are split off by an enzyme on the phage tail and the phage then detaches (TAYLOR, 1966). The well-characterised Vi antigen is thus the receptor for Vi phage attachment, although presumably some other trigger is required for injection of DNA, as this does not occur in the reversible adsorption studied *in vitro*.

## Colicin Receptors

Bacteriophage receptors seem always to be on the cell wall or cell appendages such as pili, and from the examples of cross-resistance discussed above, one would expect colicin receptors to have a similar location. However, while there is evidence for their location on the cell wall, there is also conflicting evidence for their location on the cytoplasmic membrane.

SMARDA and VRBA (1962), SMARDA (1965), and NOMURA and MAEDA (1965) find that spheroplasts are sensitive to colicins and report that colicins E2-P9, E3-CA38 and K-K235 all adsorb equally well to spheroplasts and to whole cells, suggesting that the receptor is on the membrane. Moreover, stable L-forms of *E. coli* B, which lack both the mucopeptide and outer membrane of the cell wall, retain sensitivity to colicins, including K and E2, but not to any of the T phages or phage BF23 (SMARDA, 1966 (2); SMARDA and SCHUHMANN, 1966; SMARDA and TAUBENECK, 1968]. Stable L-forms of *P. mirabilis* show similar behaviour. From this one could conclude that colicins adsorb to cytoplasmic membrane and phages to cell wall.

WELTZIEN and JESAITIS (1971), in a study discussed in more detail later, conclude that both phage T6 and colicin K receptors are on the outer membrane, although not identical. This conclusion contrasts with that of SMARDA and also with the hypothesis discussed earlier, based on the occurrence of cross-resistance, that colicin K and phage T6 adsorb to one receptor and colicins of type E and phage BF23 to another.

The reasons for these apparent discrepancies have not yet been resolved, but certain possibilities can be offered. Thus receptors may be predominantly on cell wall outer membrane but sparsely present on cytoplasmic membrane, those on cytoplasmic membrane alone being able to mediate colicin killing, and the structure of the cell wall being necessary for phage infection or for some other part of the phage growth cycle. This possibility is supported by the observation (SMARDA and TAUBENECK, 1968) that *P. mirabilis* L-forms, although very sensitive to colicin G, adsorb very little. Alternatively, colicin receptor may normally be present only on outer membrane, but in L-forms which lack outer membrane become incorporated in the cytoplasmic membrane.

BHATTACHARYYA et al. (1970) find that colicin E1 will affect proline accumulation by membrane vesicles of both sensitive strains and resistant mutants. This again indicates that the true receptor is on the cell membrane and they suggest that the cer gene "receptor" mutants have a cell wall modified so as to block access of colicin to the membrane. If so, then the existence of these mutations, conferring colicin E and phage BF23 resistance, does not imply a common receptor. However the experiments discussed below indicate that cer or tsx (T6 and colicin K resistance) mutation involves loss of an extractable receptor.

Rough mutants, with altered cell wall structure, may be sensitive to colicins which do not affect the smooth parent strain (GODARD, BEUMER-JOCHMANS and BEUMER, 1971; HAMON, 1955; MAROTEL-SHIRMAN and BARBU, 1969). This evidence is compatible with either the cell wall's being the site of the receptor or its merely shielding the membrane from colicin. MAYR-HARTING (1961, 1964) studied the effect of various agents on colicin E2 receptor function but was not able to define the chemical nature of the receptor. BORDET and BEUMER (1949, 1951) were able to extract the colicin V receptor by trichloracetic acid and found that the receptor activity resided in the same fraction as the receptor for a phage they were studying (the two do not give cross-resistance). The extracted receptor was stable at 80°, but slowly inactivated at 100°.

WELTZIEN and JESAITIS (1969, 1971) made a detailed study of the receptors of phage T6 and colicin K. The receptor activity for both resided in isolated outer membrane but not inner cytoplasmic membrane. The colicin receptor activity was labile to pepsin and pronase, but not to other proteolytic enzymes. The receptor activity was also labile to reagents which react with tryptophan and it thus appears that the receptor is proteinaceous. The T6, T2 and C16 phage receptors are also apparently proteinaceous, but the T6 and K receptor activities differ in their lability to various reagents, indicating that two different receptors are involved. REEVES (unpublished results) has found that receptor activity for colicin E2-CA42 can be retained in an extracted fraction containing mostly protein and 1% carbohydrate.

SABET and SCHNAITMAN (1971) have shown that colicin E3-CA38 and K-K235 receptor activities are associated with a high molecular weight complex (excluded by Sephadex G-200) which could be released from the outer membrane fraction of the cell envelope by use of the neutral detergent triton X-100. The E3 receptor activity was labile to trypsin, and each activity was absent in extracts from mutants resistant to the respective colicin.

It seems that although no colicin receptor has yet been purified and characterised, that they consist of protein or at least include protein.

## Conclusions and Summary

The adsorptions of colicins to specific receptors is well documented and in many ways resembles the adsorption of bacteriophages to their receptors. However, at the chemical level very little is known about colicin receptors, and less about other bacteriocin receptors, making it difficult to compare them in detail with phage receptors.

Kinetic studies of colicin killing show that cell death results from the adsorption of a single colicin molecule, but that only a proportion of adsorbed molecules do in fact kill. The average number of molecules which have to be adsorbed in order to kill a cell is known as a lethal unit.

# Chapter 6

# Mode of Action — The Biochemical Lesion

We have seen that colicins kill bacteria by first adsorbing to a receptor, but that while death of the cell seems to be effected by a single adsorbed molecule, only a small percentage of the molecules may be able to produce this effect. In this chapter we shall consider the action of colicins and other bacteriocins on the metabolism of sensitive bacteria, and for this purpose will group the bacteriocins according to their general effect.

## Bacteriocins Which Affect Energy Flux

### Colicins of Types E1 and K

JACOB et al. (1952) showed that colicin E1-ML, when added to sensitive bacteria, immediately stopped nucleic acid synthesis and growth as measured by increase in optical density, while respiration was unaffected for 20 min. LURIA (1964) showed further that this same colicin could inhibit the function of some permeases, and BHATTACHARYYA et al. (1970) showed that colicin E1-K30 inhibits permease activity of isolated membrane vesicles.

Colicin K-K235 has also been shown to inhibit nucleic acid and protein synthesis [FIELDS and LURIA, 1969 (2); NOMURA, 1963] as well as potassium transport and permease function (LURIA, 1964; NOMURA and MAEDA, 1965). Incorporation of $^{32}P$ into lipid continues at a reduced rate (NOMURA and MAEDA, 1965) and the proportions of lysophosphatidylethanolamine and diphosphatidylglycerol increase after colicin K-K235 treatment (CAVARD, POLONOVSKI and BARBU, 1967). Both colicins thus appeared to affect several functions of sensitive cells and the common denominator was thought to be some aspect of energy metabolism (LURIA, 1964). The fact that limited lipid synthesis continues shows that the effects are more specific than complete blockage of energy metabolism and FIELDS and LURIA [1969 (1, 2)] showed that neither E1-K30 nor K-K235 had much effect on the accumulation of $\alpha$-galactosides while both inhibit the accumulation of $\beta$-galactosides. They attributed the different effects to a requirement for either ATP or phosphoenol pyruvate for the accumulation phases of $\beta$-galactoside or $\alpha$-galactoside permease respectively. They suggested that the colicins lead to a lowering of ATP levels and were able to demonstrate this directly [FIELDS and LURIA, 1969 (2)].

The same two colicins do not prevent glucose being metabolised but the fate of labelled glucose is altered in colicin-treated cells such that several phosphorylated intermediates of catabolism and also pyruvate are excreted [FIELDS and LURIA, 1969 (2)]. The authors noted that the first phosphorylation of the glucose requires

ATP and that although colicin reduces the ATP level it must still be available for some reactions.

An unexplained observation is that the effects of these colicins can be mitigated by anaerobiosis [LEVINTHAL and LEVINTHAL, quoted in LURIA, 1964; FIELDS and LURIA, 1969 (2)]. However, even trace amounts of oxygen serve to sensitise cells to colicin without altering the overall fermentative pathway.

It seems likely that the colicins do affect energy metabolism although it is not yet possible to define their effect at the molecular level. FIELDS and LURIA [1969 (2)] suggest possible pathways for the observed effects of colicin on the hypothesis that reactions involving ATP are involved in the primary lesion. An alternative hypothesis, that the manifold effects are due to gross membrane damage with subsequent general leakage of low molecular weight compounds, is not supported as not only are $\alpha$-galactosides accumulated [FIELDS and LURIA, 1969 (1)] but $\beta$-galactosides still require the specific permease to facilitate their diffusion across the membrane (LURIA, 1964).

Several other colicins of types E1 and K (E1-N104, K-K40, K-K235, K-K297 and K-K319) all have a similar effect on *E. coli* K12, in that both DNA and protein synthesis are inhibited very rapidly by excess colicin (REEVES, 1968). This is in contrast to the effects of the other colicins to be discussed, which even in excess have only a delayed effect on one or the other of these two. Under the conditions used all five had a more marked effect on protein than on DNA synthesis if lower amounts of colicin were used.

Colicin E1-ML has been shown to stop multiplication of a phage (JACOB *et al.*, 1952) and K-K235 has the same effect on phages BF23, T2 and T6 [FREDERICQ, 1953 (2)] and on phage T4 in *E. coli* B (NOMURA, 1963). In the latter case the colicin is as effective in destroying infected cells assayed as plaque forming units as it is in killing uninfected cells.

## Colicins of Types A and I

Colicins Ib-P9 and Ia-CA53 both inhibit DNA, RNA and protein synthesis, while having less effect on synthesis of phospholipids and low molecular weight organic phosphates, and no effect on respiration (LEVISOHN *et al.*, 1968). It seems that these two colicins act similarly to colicins of types E1 and K. A colicin of type A has been shown to act similarly in many respects (NAGEL DE ZWAIG, 1969). It inhibits nucleic acid synthesis and accumulation of isoleucine and $\beta$-galactoside. It did not affect accumulation of an $\alpha$-galactoside. Likewise colicin A-CA31 inhibits protein, RNA and DNA synthesis of a sensitive *Klebsiella edwardsii* strain (DE GRAAF and STOUTHAMER, 1971).

## Pyocin P10

In 1954 JACOB, using the pyocin produced by *Pseudomonas pyocyanea* P10, showed that addition of pyocin to a sensitive culture stopped growth, measured as optical density, and the respiration rate began to fall instead of rising steadily. Both of these observations suggest that this pyocin may act in the same way as colicins E1-ML and K-K235. This is borne out by the further observation that the pyocin can stop multiplication of a virulent bacteriophage in a sensitive cell. This pyocin perhaps acts on the utilization of energy in a way similar to that in which colicins of types E1 and K act.

A material produced by *Pseudomonas aeruginosa* R and referred to as a pyocin, but more probably a defective bacteriophage (see Chap. 2), perhaps has a similar mode of action [KAZIRO and TANAKA, 1965 (1, 2; KONISKY and NOMURA, 1967]. The continuing incorporation (at reduced rate) of $^{32}$P into nucleotides suggests that the effects are not due to breakdown of the permeability barrier as observed for other bacteriophages (see discussion later).

# Bacteriocins Which Affect DNA Metabolism

## Colicins of Type E2

Colicins of type E2 all seem to act on DNA metabolism, although most of the evidence relates to colicins E2-P9 (HOLLAND and HOLLAND, 1970; NOMURA, 1963, 1964; NOMURA and MAEDA, 1965) and E2-CA42 (REEVES, 1966; REYNOLDS and REEVES, 1963, 1969).

These two colicins are similar in their mode of action but, although both of type E2, differ in their antigenic specificity (LEWIS and STOCKER, 1965). Their different activity spectra led FREDERICQ (1948) to first classify them as colicins of types F (CA42) and S3 (P9).

Excess of either colicin added to a culture of *E. coli* K12 stops DNA synthesis within 2—5 min. RNA synthesis stops next and protein synthesis stops at about 15 min. Respiration is unaffected for about 20 to 30 min and then drops.

Not only is DNA synthesis inhibited but preexisting DNA is degraded. Acid soluble degradation products are detected about 5 min after treatment with either colicin and continue to accumulate for about 30 min. The proportion of the DNA degraded and the kinetics of degradation depend on the multiplicity of colicin used (HOLLAND and HOLLAND, 1970; NOMURA and MAEDA, 1965; REYNOLDS and REEVES, unpublished data) and on the sensitive strain (HULL and REEVES, 1971). It has recently been shown that the degradation of DNA to acid soluble material is preceded by endonucleolytic cleavage of the DNA into large fragments with a molecular weight of about $10^6$ (HULL and REEVES, unpublished data; OBINATA and MIZUNO, 1970; RINGROSE, 1970).

Single strand breaks have been detected in the DNA of colicin treated cells after 2 min, before double strand breaks were detected (RINGROSE, 1970). Also the different molecular weights of DNA, as determined on alkaline and neutral sucrose gradients, suggested that even after double strand breaks have occurred some single strand breaks still exist. Presumably the effect of colicin is to activate an endonuclease which gives single strand breaks, and these are then converted to double-strand breaks and the DNA further degraded by exonucleolytic action.

Colicin E2-P9 has no detectable nuclease activity itself (NOMURA, 1964; RINGROSE, 1970), although this lack of *in vitro* activity may be due to the presence of an inhibitor in the colicin preparations. As we shall see colicin E3-CA38 has an effect *in vitro*, but crude colicin preparations contain an inhibitor (BOWMAN et al., 1971) of this effect.

However, so far it has generally been thought that colicin E2 activates a host nuclease, and it is not clear which nuclease could be involved. Strains lacking endonuclease I degrade DNA normally after colicin action (HULL and REEVES, unpublished data; OBINATA and MIZUNO, 1970), although LURIA and SAXE (1971) report otherwise with respect to λ DNA (see below).

UvrA, UvrB and UvrC mutants behave similarly to wild type *E. coli* K12 after colicin addition, in that their DNA is degraded with similar kinetics (HULL and REEVES, unpublished data). This suggests that the excision enzymes, which initiate degradation induced by ultraviolet irradiation, are not involved in colicin induced degradation of DNA.

After colicin treatment DNA is degraded more rapidly by RecA mutants, and more slowly by RecB and RecC mutants, than by isogenic Rec+ strains (HULL and REEVES, unpublished data). This resembles the kinetics of DNA degradation in these strains after ultraviolet irradiation (WILLETTS and CLARK, 1969). Presumably the ATP dependent nuclease present only in RecB+ RecC+ strains (BARBOUR and CLARK, 1970) is involved in colicin and radiation induced degradation, but since degradation occurs also in its absence, even if more slowly, this enzyme cannot be involved in the primary effect.

The degradation does not appear to occur specifically at any point on the chromosome as radioactivity incorporated near the replicating fork or near the origin of replication, becomes acid soluble with normal kinetics (BEPPU and ARIMA, 1971).

Both colicin E2-P9 and E2-CA42 will induce strains of *E. coli* K12 lysogenic for phage λ (ENDO, KAMIYA and ISHIZAWA, 1963; NOMURA, 1963; REEVES, unpublished data). At low multiplicities of E2-P9 (and also E2-CA42) the induced cells (which die on lysis) account for all the bacteria killed and their number is proportional to the amount of colicin used. At higher multiplicities of colicin the proportion of the killed cells which release phage decreases.

Both colicins have been shown to have little effect on virulent bacteriophages infecting sensitive cells. NOMURA (1963) showed that bacteriophages T4 and T5 can infect and replicate within cells which have been 25 min beforehand "killed" with colicin E2-P9, and also that the DNA of vegetative T4 bacteriophage is degraded only slowly. Indeed the degradation was only detected if the infected bacteria were simultaneously treated with chloramphenicol. Colicin E2-CA42 can however cause substantial degradation of labelled λ DNA, regardless of whether the label is present in replicating DNA or in the nonreplicating form present after superinfection of lysogenic bacteria (HULL and REEVES, 1971; SAXE and LURIA, 1971). However, after λ infection, both host and phage DNA are degraded more slowly than DNA of uninfected bacteria, and degradation is more easily inhibited by starvation (HULL and REEVES, 1971).

SAXE and LURIA (1971) made the very interesting observation that after colicin treatment, mutants lacking endonuclease I did not degrade λ DNA to acid soluble material, although supercoils were converted to linear λ DNA. This contrasts with the earlier observation that a similar mutant did degrade *E. coli* DNA after colicin treatment.

HOLLAND and HOLLAND (1970) report that λ DNA is not degraded in E2 treated cells. The apparent discrepancy is probably due to use of different amounts of colicin (HULL and REEVES, 1971).

Other work on the action of E2-P9 suggests that its effect may not be confined to DNA as there is extensive degradation of RNA beginning 30—60 min after

addition of colicin (NOSE and MIZUNO, 1968; NOSE, MIZUNO and OZEKI, 1966). The
long delay before any observed effect however makes it readily conceivable this is a
secondary effect. Perhaps more significant for an understanding of the primary effect
is the report of FUJIMURA (1966) who finds that this same colicin can inhibit replication of the RNA of the R17 bacteriophage. This RNA is of course being replicated
by an RNA dependent RNA polymerase yet its synthesis seems to be affected in the
same way as the normal cellular RNA synthesis, after delay of 5—10 min. This
suggests that the effect of colicin E2-P9 on RNA synthesis is direct, or if
secondary to the effect on DNA synthesis, is at least not simply explained by
template destruction.

Although much remains to be learnt about E2 action, it now seems clear that the
primary biochemical lesion is the cleavage of the DNA at intervals of approximately
$10^6$ molecular weight along the chromosome. If the breaks occur at specific sites on
the DNA, then the observed differences in the rate of degradation of host, T4 and $\lambda$
phage DNA may reflect differences in the distribution of such sites. It is of interest
that cytosine rich clusters occur at intervals of $1 - 2 \times 10^6$ molecular weight along
the DNA of *E. coli* and $\lambda$, but are far less frequent in T4 DNA (SZYBALSKI,
KUBINSKI and SHELDRICK, 1966). HULL and REEVES (1971) speculated on the basis
of this correlation that cytosine rich clusters may constitute the specific sites cleaved
in colicin E2 action.

The endonuclease involved is not yet known although we have seen that there is
evidence for and against endonuclease I being involved. Another possibility is endonuclease II which fragments *E. coli* DNA to pieces of about $10^6$ molecular weight
and, by analogy with the very similar T4 endonuclease II, probably acts specifically
at cytosine rich clusters (FRIEDBERG, HADI and GOLDTHWAITE, 1969; KUTTER and
WIBERG, 1968). Endonuclease II has already been implicated in replication (HASKELL
and DAVERN, 1969) and recombination (FRIEDBERG and GOLDTHWAITE, 1969).

Other colicins of type E2 have been studied in less detail. Colicin E2-K317
affects macromolecular synthesis as do E2-CA42 and E2-P9, and it induces phage P1
(ELGAT and BEN GURION, 1969). Colicins E2-GEI288, E2-GEI554, E2-GEI602 and
E2-M5092 all behave like the other two in having a much more rapid effect on DNA
than on protein synthesis, and even at very high colicin levels, when DNA synthesis
is rapidly stopped, protein synthesis continues at a reduced rate for about 15 min
(REEVES, 1968). Thus all colicins of type E2 may have the same mode of action.

Colicin P can also induce prophage and hence by analogy may perhaps affect DNA
synthesis [HAMON and PERON, 1965 (2)].


## Megacins C-C4 and KP-337

Megacin C-C4, produced by *B. megaterium* C4 appears to act similarly to the two
colicins we have been discussing (HOLLAND, 1963, 1965). On addition to growing
cultures of *B. megaterium* MUT it stopped DNA synthesis and greatly reduced protein
and RNA synthesis, and also led to degradation of DNA. Unlike the colicins this
megacin does not completely inhibit RNA or protein synthesis, even after 90 min.
In fact, the continued synthesis of RNA and protein at 90 min when 50% of the
DNA had been degraded by 30 min is of considerable interest.

A second megacin, KP-337 produced by *B. megaterium* 337, resembles megacin C in several respects (see Appendix) but the studies on its mode of action (DURNER and MACH, 1966) do not yet permit a detailed comparison as only the effect of a high concentration of megacin was studied. This had a marked effect on DNA, RNA and protein synthesis by the time the first sample was taken at 15 min. Megacin KP-337 also causes the cells to leak all their constituents by 60 min but again only a very high concentration was used and at present the primary effect of this megacin is not known.

### Pesticin

Pesticins of type P1 (see Appendix) will act on *E. coli* and have similar effects to colicins of type E2. DNA synthesis is stopped immediately whereas RNA and protein synthesis continue for about 90 and 60 min respectively. The pesticin also induced prophage in *E. coli* (ELGAT and BEN GURION, 1969).

## Bacteriocins Which Affect Protein Synthesis

### Colicins of Type E3

Colicin E3-CA38 has recently been shown to act *in vitro* [BOWMAN *et al.*, 1971 (2)]. The discovery of an *in vitro* effect of a colicin marks a major step forward in understanding colicin action and is particularly significant in that it was largely unexpected.

Colicin E3-CA38 has been known for some time to affect protein synthesis. At high colicin concentrations (multiplicity of infection of 96) synthesis stops within a few minutes, but at lower multiplicities it takes longer to stop it, and at a multiplicity of infection of 3.5 protein synthesis is still continuing slowly after 80 min (NOMURA, 1964). Colicin E3 is said to have no effect on DNA synthesis (NOMURA, 1963) and has no detectable effect on overall RNA synthesis (NOMURA, 1963; SENIOR, KWASNIAK and HOLLAND, 1970) but does induce a short-lived increase in mRNA stability or rate of synthesis (SENIOR *et al.*, 1970).

KONISKI and NOMURA (1967) have found that after adsorption of E3-CA38, the ribosomes themselves become altered and cell extracts unable to perform *in vitro* protein synthesis. The defect has been localised in the RNA component of the 30S subunit of the ribosome (BOWMAN, DAHLBERG, IKEMURA, KONISKY and NOMURA, 1971; SENIOR and HOLLAND, 1971). Both groups were unable to find any effect on the acrylamide gel electrophoresis pattern of the proteins of the 30S subunit, and BOWMAN *et al.* [1971 (1)] showed that the proteins were functional if removed and reconstituted with RNA from unaffected ribosomes. However, the RNA from 30S ribosomal subunits of colicin-damaged cells differs from the RNA from untreated cells in having a reduced sedimentation coefficient (SENIOR and HOLLAND, 1971), and BOWMAN *et al.* [1971 (1)] isolated a fragment derived from the 3' end. Colicin E3 treatment thus causes a cleavage of the 30S ribosomal subunit RNA about 50 nucleotides from the 3'-terminus.

This discovery was remarkable enough, indicating that the action of colicin E3-CA38 resulted in a specific break in the ribosomal RNA, and was quickly followed by the further discovery that this colicin had the same effect on ribosomes *in vitro* [Boon, 1971; Bowman *et al.*, 1971 (2)]. This significant step in the elucidation of the mode of action of colicins came at the time this monograph was going to press, and hence is not taken into account as fully as it might be elsewhere in the text. The *in vivo* and *in vitro* effects appear to be identical and presumably the colicin, after adsorption to the receptor, enters the cell and affects ribosomes directly, in marked contrast to most earlier hypotheses on colicin action. The *in vitro* studies indicate that the effect of colicin is catalytic and that the colicin is either an RNAase or activates a ribosomal RNAase. This discovery will lead to a renewed search for *in vitro* activity of other colicins, and a better understanding of colicin action in general.

Ribosomes from colicin producing cells were sensitive to colicin [Boon, 1971; Bowman *et al.*, 1971 (2)] and the immunity of the colicinogenic strain was shown to be due to an intracellular inhibitor [Bowman *et al.*, 1971 (2)]. This inhibitor is present in crude colicin preparations, and residual inhibitor in colicin preparations probably accounted for the failure of earlier attempts to demonstrate an *in vitro* effect of colicin E3-CA38 (Hull, 1971; Konisky, 1967).

Reeves (1968) found colicin E3-K365 to also protein affect synthesis before DNA synthesis. This result for the only other colicin of type E3 tested suggests that, as for the other colicins of type E, mode of action correlates well with subdivision into subtypes E1, E2 and E3.

### Cloacin DF13 and Pneumocins S6 and S8

De Graaf and Stouthamer (1969) and de Graaf *et al.* (1969) found that cloacin DF13 resembles colicin of type E3 in that it inhibits protein synthesis while DNA and RNA synthesis and respiration continue normally. De Graaf and Stouthamer (1971) made similar observations for pneumocins S6 and S8, although the inhibition of protein synthesis was delayed about 20 min.

## A Bacteriocin Affecting Protein and RNA Synthesis

### Pneumocin G196

A single pneumocin inhibits protein and RNA synthesis after a delay of 15 min, while having no effect on DNA synthesis (de Graaf and Stouthamer, 1971). This bacteriocin thus seems to be unique in its effect.

## Mediation of the Bacteriocin Effect

We have seen that the bacteriocin molecule which adsorbs to the surface receptor affects cell metabolism in one of three ways, inhibiting either protein or DNA synthesis or some aspect of energy metabolism.

As we saw in Chap. 5, cells which have been inhibited by colicin K-K235 for 25 min can be rescued by trypsin (Nomura and Nakamura, 1962). In this case the continued presence of the colicin on the surface is required for lethality and presumably the effect is mediated continuously. In the case of colicins of type E2 or pesticin P1, the effects are rapidly irreversible, but in the presence of 2:4 dinitro-phenol (DNP) (REYNOLDS and REEVES, 1963) or cyanide (HULL and REEVES, un-published data) the cells are rescuable for 1 or 2 h after colicin E2-CA42 adsorption. How is the effect of the adsorbed bacteriocin exerted?

Until the demonstration that colicin E3-CA38 could directly affect ribosomes [BOWMAN et al., 1971 (2)] it was generally assumed that colicins remained attached to a surface receptor, while exerting a lethal effect on the cell. Since colicin E3-CA38, or at least a part of it, must surely enter the cell, the possibility arises that other coli-cins act in this way, although as yet there is no evidence for this.

## Immunity of Colicinogenic Bacteria

Bacteria are generally immune to any bacteriocin they produce [FREDERICQ, 1954 (1)] although they retain the ability to adsorb it as demonstrated for colicin E2-P9 (MAEDA and NOMURA, 1966). Thus these immune cells do not respond to adsorbed colicin in the same way and some step in colicin-mediated death must be modified. NOMURA (1963) suggested that in E. coli K12 (E2-P9) it is not the DNA itself which is modified as for instance occurs in host modification systems (see ARBER and LINN, 1969, for review), because DNA transferred from a sensitive donor to a colicinogenic recipient immediately became immune to the effect of added colicin E2-P9. He also studied the development of immunity in cells to which the C-factor E2-P9 had just been transferred and found that this development required reasonable growth conditions, occurring only slowly in buffer without nutrients and not at all in the cold, but he could demonstrate no absolute requirement for DNA or protein synthesis.

The nature of immunity to colicins is still generally unknown, but in the case of colicin E3-CA38, has recently been shown to be due to an inhibitor [BOWMAN et al., 1971 (2)]. The inhibitor has yet to be purified, and the details of its action are un-known, but as it does not irreversibly affect either colicin or the target ribosomes, it presumably inhibits the interaction between colicin and ribosomes.

In all other cases we can conclude only that the immune system must interact in some way with the mediation of the colicin effect.

## The Effect of Specific Inhibitors on the Effect of Bacteriocins

If the cell is prevented from either mobilizing energy or synthesizing protein, RNA or DNA by some specific means, then we might hope to show whether these particular activities are required for the effect of bacteriocin to be exerted. After treatment of the bacteria with bacteriocin in the presence of the inhibitor, we can see whether the bacteria have been irreversibly damaged by removing the bacteriocin with trypsin and assaying for viability. In some cases it is also possible to see whether the colicin is affecting cellular metabolism in the presence of the other inhibitor.

## Is Energy Necessary?

Whereas trypsin rescue is normally effective for only about 5 min after colicin E2 addition, rescue can be effective for up to 2 h if the cells are treated with DNP

(REYNOLDS and REEVES, 1963) or cyanide (HULL and REEVES, unpublished data). Both agents and also colicin K will prevent colicins of type E2 from inducing DNA degradation to acid-soluble products (HULL and REEVES, unpublished data; NOMURA and MAEDA, 1965). All three agents are thought to affect energy metabolism, although as we have seen, the details of K-K235 action are not understood and DNP is used at high concentrations ($2 \times 10^{-3}$ M) and may be having extensive side effects. Both DNP and cyanide will reduce the rate of DNA degradation if added after it has commenced (HOLLAND and HOLLAND, 1970), and hence the energy requirement may be for the degradation itself rather than for intermediate steps.

However, if colicins of type E2 resemble colicin E3-CA38, and enter the cell, then energy may also be required for colicin entry, or alternatively colicin which enters in the presence of inhibitors such as cyanide, is unable to have any effect even after the inhibitor is removed.

Although cyanide can prevent degradation of DNA, deprivation of energy by bubbling nitrogen through lactate-grown bacteria does not stop degradation proceeding normally after colicin addition (REYNOLDS, 1966). It would seem either that the energy requirements for degradation are very low, or, less probably, that cyanide is affecting colicin action directly and not through its effect on electron transport.

In contrast to the observations with colicins of type E2, megacin C-C4 cannot initiate DNA degradation in *B. megaterium* (a strict aerobe) if the energy supply is cut off by cessation of aeration (HOLLAND, 1965).

## Is Protein or RNA Synthesis Necessary?

In the case of colicin E2-CA42 and megacin C-C4, pretreatment with chloramphenicol (CM) reduces but does not completely prevent the DNA degradation consequent to addition of the bacteriocin (HOLLAND, 1965; REYNOLDS, 1966) and CM is reported to have no effect on E2-P9 provoked DNA degradation (HOLLAND and HOLLAND, 1970; NOMURA, 1963). HOLLAND also observed that streptomycin could only partially prevent megacin DNA degradation. It thus seems that protein synthesis is not necessary for the bacteriocin effect to be mediated.

## Is DNA Synthesis Necessary?

REYNOLDS (1966) showed that pretreatment with nalidixic acid, which inhibits DNA synthesis (COOK, DEITZ and GOSS, 1966; DEITZ, COOK and GOSS, 1966) would prevent colicin E2-CA42 caused DNA degradation. In contrast to this NOMURA and MAEDA (1965) found that thymine deprivation of a Thy- strain does not prevent such degradation. It would seem that the synthesis of DNA is not necessary for the mediation of the colicin effect but that the biochemical target of nalidixic acid, perhaps the DNA polymerizing system (GOSS, DEITZ and COOK, 1965), must be functional if colicin E2-CA42 is to initiate DNA degradation.

## Colicin-Tolerant Mutants

Another experimental approach to the problem of mediation of the colicin effect involves the use of mutants which, while still able to adsorb a colicin, are resistant

Table 6.1. *Colicin-tolerant mutants*

Some properties of P109, a substrate of *E. coli* K 12 and 10 mutants derived from it, being one typical Cer mutant (P125), one (P120) which is probably an atypical Cer mutant and 8 Tol mutants which map near Gal. After Reeves (1966)

| | P109 | P114 P118 P134 P138 | P117 P119 | P116 | P137 | P120 | P125 |
|---|---|---|---|---|---|---|---|
| **Reaction to colicins** colicin E(a)[a] | s | t | t | t | t | t | t |
| colicin E(b)[a] | s | t | t | s | s | s | t |
| colicin E(c)[a] | s | t | s | s | s | t | t |
| K-K235 | s | t | s | s | s | t | t |
| A-CA31 | s | t | t | t | s | s | s |
| C-CA57 | s | t | s | s | s | s | s |
| E.o.p. of BF23 | 1.0 | .3 | .3 | .4 | .5 | $<10^{-9}$ | $<10^{-9}$ |
| Linkage of colicin resistance locus | – | Gal | Gal | Gal | Gal | Arg | Arg |
| Adsorption of colicin E2-CA42 | + | + | + | + | + | – | – |
| Most similar class in Table 6.2 | | II | IV | | | | |

[a] Group a comprised 13 colicins of type E2, group b comprised 5 colicins of type E2 and 2 of type E3, while group c comprised 4 colicins of type E1 and perhaps one of type E2.

to it. The mutations presumably affect either the mediation of the colicin effect or the biochemical target so as to nullify the effect of colicin. Such mutants are said to be tolerant (or refractory) to colicin. Mutants tolerant to colicins of type E were introduced in Chap. 1, and are recognised as retaining the receptor because they

Table 6.2. *Some of the properties of the colicin-tolerant mutants isolated from E. coli K12 strains*

From HILL and HOLLAND (1967), NAGEL DE ZWAIG and LURIA (1967) and NOMURA and WITTEN (1967). Classification is that of the latter two papers.

| | Mutant Class | | | | | | | | |
|---|---|---|---|---|---|---|---|---|---|
| | I[c] | II[a,b,c] | III[a,b,c] | IV[a,c] | V[a,c] | VI[c] | VII[a,c] | VIII[a,b] | —[a] |
| Reaction to colicins E1-K30 | s | t | s | s | t | t | s | t | s |
| E2-P9 | s | t | t | t | t | t | t | s | s |
| E3-CA38 | s | t | t | t | t | s | s | s | t |
| K-K235 | t | t | t | s | s | s | s | s | t |
| A[d] | — | — | — | — | — | — | — | s | — |
| Map location near | Gal | Gal[e] | Gal[e] | Gal[e] | Gal[e] | — | SerB | Thy | — |
| Classification of HILL and HOLLAND | — | VI,VIII | V,VII | IV,VII | VI | — | II | I | III |

a Classes described by HILL and HOLLAND.
b Classes described by NAGEL DE ZWAIG and LURIA.
c Classes described by NOMURA and WITTEN.
d Tested only by NAGEL DE ZWAIG and LURIA.
e Some mutants not linked to Gal.

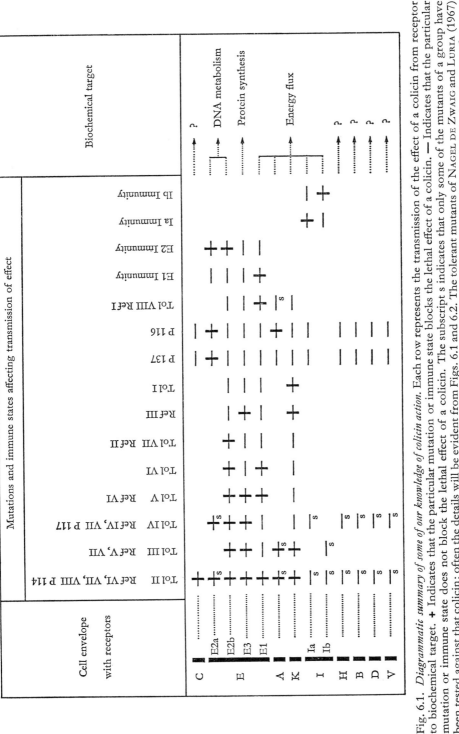

Fig. 6.1. *Diagrammatic summary of some of our knowledge of colicin action*. Each row represents the transmission of the effect of a colicin from receptor to biochemical target. **+** Indicates that the particular mutation or immune state blocks the lethal effect of a colicin. — Indicates that the particular mutation or immune state does not block the lethal effect of a colicin. The subscript s indicates that only some of the mutants of a group have been tested against that colicin; often the details will be evident from Figs. 6.1 and 6.2. The tolerant mutants of NAGEL DE ZWAIG and LURIA (1967) and NOMURA and WITTEN (1967) are labelled Tol, these of HILL and HOLLAND (1967), Ref, and one example is included of each type of mutant described by REEVES (1966)

retain sensitivity either to some colicins of type E or to phage BF23, thought to adsorb to the same receptor. Mutants with one or the other or both of these properties have been described several times (CLOWES, 1965; FREDERICQ, 1948; HILL and HOLLAND, 1967; HOLLAND, 1967, 1968; HOLLAND and THRELFALL, 1969; NAGEL DE ZWAIG and LURIA, 1967, 1969; NOMURA, 1964; NOMURA and MAEDA, 1965; NOMURA and WITTEN, 1967; RAMPINI, SCHIRMAN and BARBU, 1967; REEVES, 1966).

The earlier papers did no more than note the existence of these mutants but the later ones recognise considerable heterogenity amongst them, and it is difficult to equate the classifications used by the various authors.

The observation that many of the mutants are also tolerant to colicins of types A, C or K suggests that colicins adsorbing to different receptors may mediate their effects via common cellular components (NAGEL DE ZWAIG and LURIA, 1967; RAMPINI et al., 1967; REEVES, 1966).

Tolerant mutants are divided into various classes. REEVES (1966) recognised five classes on the basis of their tolerance to many colicins of type E and single colicins of other types. NAGEL DE ZWAIG and LURIA (1967) and NOMURA and WITTEN (1967) recognise eight classes on the basis of tolerance to five colicins. Some of the properties of the classes of mutants are shown in Tables 6.1 and 6.2, and again summarised in Fig. 6.1. The class designations Tol I-VIII are being replaced with genetic symbols as the genes are characterised.

## The Tol VIII Class or TolC Gene

CLOWES (1965), HILL and HOLLAND (1967) and NAGEL DE ZWAIG and LURIA (1967, 1969) all describe a class of mutants originally known as Tol VIII. They have recently been shown to map at 59 min by WHITNEY (1971) who also did fine structure mapping. These TolC mutants are tolerant only to E1-K30 (not E2-P9, E3-CA38, A-CA31 or K-K235) and are also highly sensitive to various dyes and detergents. NAGEL DE ZWAIG and LURIA (1967) have demonstrated an increased uptake of dyes by the mutants, which is not prevented by cyanide, suggesting that active transport is not involved in the uptake. They consider that the mutation involves membrane damage which permits leakage of the dyes inwards. ROLFE and ONODERA (1971) have confirmed this by showing that the membrane of a TolC deletion mutant lacks a membrane protein. NAGEL DE ZWAIG and LURIA (1967) propose that this locus is involved in the transmission of the colicin effect, which they consider may well be via membrane. On the other hand, we cannot in this case eliminate the possibility that the mutation affects the biochemical target of E1, perhaps a membrane-associated step in energy metabolism, in which case the existence of these mutations would offer no evidence on the nature of the transmission of the colicin effect. The Tol VIII gene product must be present on membrane vesicles since these also are tolerant to the effect of colicin E1 on permease action (BHATTACHARYYA et al., 1970).

NAGEL DE ZWAIG and LURIA (1969) obtained a Tol VIII mutant which in the presence of both a Str and an amber suppressor SuII mutation was tolerant at 41° but sensitive a 30°. The temperature-sensitive phenotype presumably depends on a Tol VIII protein with changed temperature stability. They observed that after a temperature shift, the appearance of either tolerance or sensitivity was dependent on de novo protein synthesis (Table 6.3).

The existence of this mutant tells us that the Tol VIII gene product is a protein. Since CM can prevent the change from one phenotype to the other after a temperature shift, it seems that after synthesis and perhaps integration into some cell structure, the mutant protein cannot be converted from one form to the other by temperature change alone.

### The Tol VII or Ref II Gene

Another class of mutants are tolerant only to colicin E2-P9 (not E1-K30, E3-CA38 or K-K235) and map at about 90 min near Thr on the K12 genetic map. These mutants are known as Ref II (refractory) (HILL and HOLLAND, 1967; HOLLAND, 1967, 1968; HOLLAND and THRELFALL, 1969) or CetA, CetB and CetC (HOLLAND, THRELFALL, HOLLAND, DARBY and SAMSON, 1970), but are probably the same as the Tol VII mutants (NOMURA and WITTEN, 1967) which have similar tolerance properties but have not been mapped genetically. These mutants have been divided into three groups, CetA, B and C (HOLLAND et al., 1970; THRELFALL and HOLLAND, 1970), all of which map at about 90 min.

The tolerance of CetC mutants may be exhibited at low temperature only (25°); the mutants may also show defective division, recombination or λ propagation, and may be sensitive to detergents. All CetC mutants are sensitive to ultraviolet radiation and the various properties of a CetC mutant are co-transducible and apparently due to a single mutation (THRELFALL and HOLLAND, 1970). CetB mutants are not pleiotropic and are tolerant only at low temperature, while the CetA locus in some way affects expression of CetB.

Further analysis of the CetC defects in recombination and U.V. sensitivity may well reveal the nature of the colicin E2 induced lesion.

### The Tol I to VI Mutants

Another group of mutations, many mapping at the TolA and TolB loci near Gal at 17 min, almost certainly affect relatively early steps in the mediation system, as they may confer tolerance to colicins which inhibit such apparently unrelated functions as DNA synthesis or protein synthesis.

The mutants at these loci are variable in their phenotype, and are subdivided into various classes (now known as Tol I—VI) on the basis of their variable sensitivity patterns to colicin of types A, C, E1, E2, E3 and K (HILL and HOLLAND, 1967; NAGEL DE ZWAIG and LURIA, 1967; NOMURA, 1964; NOMURA and WITTEN, 1967; RAMPINI et al., 1967; REEVES, 1966). REEVES (1966) found a considerable correlation between the subdivision of colicins into sub-types E1, E2 and E3 and their action on the various classes of tolerant mutants (Table 6.1). Three groups (NAGEL DE ZWAIG and LURIA, 1967; NOMURA and WITTEN, 1967; REEVES, unpublished data) have studied the dominance of three mutations mapping near Gal using an F' Gal epichromosome carrying the wild type Tol allele. The wild type was dominant in all cases.

There are at least two Tol loci near Gal since Tol II and Tol III mutants will complement (NAGEL DE ZWAIG and LURIA, 1967) and there are other Tol loci not linked to Gal in addition to TolC at 59 min and Ref II near Thr (HILL and HOLLAND, 1967; NAGEL DE ZWAIG and LURIA, 1967; NOMURA and WITTEN, 1967).

Table 6.3. *Properties of temperature-sensitive tolerant mutants*

| Class | Mutant used in biochemical studies | Reaction to colicins in next column | | Colicins to which tolerance is temperature-dependent | Colicins to which tolerance is temperature-independent | Characteristics of tolerance on temperature change | | References |
|---|---|---|---|---|---|---|---|---|
| | | 25 or 30° | 40 or 41° | | | from tolerant to sensitive state | from sensitive to tolerant state | |
| Tol II | LA 639 | S | T | A, E1, E2, E3, K | — | CM sensitive | CM sensitive | NAGEL DE ZWAIG and LURIA (1969) |
| RefVI (= Tol II?) | | S | T | E1 | E2, E3, K | — | — | HILL and HOLLAND (1967) |
| RefVII (= Tol III?) | | S | T | E3 | E2, K | CM sensitive | Not CM sensitive | |
| Tol IV | ER 437 | S | T | E2, E3 | | CM sensitive | — | NOMURA and WITTEN (1967) |
| | ER 502-24, ER 502-34 | S | T (no growth at 40°) | E2, E3 | | — | — | |
| Tol VII | ER 438 | T | S | E2 | | Not CM sensitive Immediate | Not CM sensitive Immediate | HOLLAND (1968) |
| Ref II (= Tol VII?) | | T | S | E2 | | | | HILL and HOLLAND (1967) |
| Ref III | | S | T | E3 | | | | |
| Tol VIII | LA 641 | S | T | E1 | | CM sensitive | CM sensitive | NAGEL DE ZWAIG and LURIA (1969) |

The tolerance of some of these mutants is expressed only at high temperature (40°) (HILL and HOLLAND, 1967; HOLLAND, 1968; NAGEL DE ZWAIG and LURIA, 1969; NOMURA and WITTEN, 1967). The sensitivity of these temperature-dependent mutants appears to be normal at 30° in that colicin E2-P9 causes DNA degradation and colicin E3-CA38 stops protein synthesis (NOMURA and WITTEN, 1967) (Table 6.3).

The Tol II mutant LA639 (NAGEL DE ZWAIG and LURIA, 1969) and the Tol IV mutant ER437 (NOMURA and WITTEN, 1967) both require protein synthesis after the temperature is lowered for colicin sensitivity to be expressed, but after the converse change only LA639 requires protein synthesis for expression of tolerance.

NAGEL DE ZWAIG and LURIA (1969) studied the kinetics of these conversions in LA639 (Tol II) and found that colicin E1-K30, E2-P9 or E3-CA38 adsorbed at 41° could not kill when the temperature was subsequently lowered to 30°, suggesting that those receptors synthesized at 41° are unable to participate in lethal events even after transfer to 30°. The kinetics with which LA639 became sensitive to various colicins after shifting from 41° to 30° suggest that there is a simple dilution of colicin-tolerant sites with colicin-sensitive sites. After the converse shift it seems, however, that some of the sensitive sites synthesised at 30° became unresponsive to colicins at 41°.

Two mutants, ER502-24 and ER502-34, which map around 65 min resemble the other Tol IV conditional mutants except that they cannot grow at 40° (NOMURA and WITTEN, 1967). However, they are tolerant at 40° in that colicins have no biochemical effect at this temperature. However, in contrast to LA639, colicin adsorbed to these mutants at 40° could kill if the temperature was subsequently changed to 30°. Similar mutants are described by ROLFE, BERNSTEIN, ONODERA and BECKER (1971). Other mutants, probably Tol II, are tolerant at 30° but do not grow at 40° (RAMPINI et al., 1967). The basis of their temperature sensitivity could not be determined. In both cases we have a mutation conferring colicin tolerance and simultaneously affecting the viability of the cell.

Deletion mutants have been reported which encompass the whole of the Tol region near Gal and hence none of the proteins encoded by these genes can be essential for viability (ONODERA, ROLFE and BERNSTEIN, 1970). Recently the same group (BERNSTEIN, ROLFE, ONODERA and TILL, 1971) started a fine structure genetic analysis.

## Cross-Resistance between Colicins B, I and V

Another group of colicins show frequent cross-resistance and these mutants are mentioned here to compare them with Tol mutants which show cross-resistance between colicins A, C, E and K. PAPAVASSILIOU [1960 (3)] found frequent cross-resistance between colicins B, I and V and GRATIA (1964) found that mutants resistant to one or more of the colicins B-CA18, I-CA53, V-CA7 frequently show resistance to phages T5 and T1 and various host range mutants of T1. The resistance spectra of the various mutants (Table 6.4) show that if the agents are arranged in the order T5, T1 (and its mutants), B, I and V, that all mutants resist a group of agents contiguous in the table. All but the first three classes of mutants in Table 6.4 map near to Try, and are probably deletions encompassing one or more genes. Some of the deletions extend into the Try operon. Although the nature of the resistance is not known, the results seem compatible with the existence of several closely linked genes, each determining sensitivity or resistance to one of the colicins involved. The significance of this close linkage is at present unknown, but in the case of colicin B, the resistance is in part due to secretion of enterochelin, a material that neutralises colicin (GUTTERMAN, 1971; GUTERMAN and LURIA, 1969).

Table 6.4. *Spectrum of resistance of single-step mutants of E. coli selected by colicins B, I or V or phage T1 or all four agents simultaneously. Some details of resistance to T1 host range mutants have been omitted. (From* GRATIA, *1964)*

| Agent used to select mutant | Resistance to | | | | | | |
|---|---|---|---|---|---|---|---|
| | T5 | T1h$_1$ | T1 | T1h$_2$ | B | I | V |
| T1 | + | + | + | + | ± | | |
| T1 | | | + | + | ± | | |
| T1 | | + | + | | | | |
| T1, B, I and V | | + | + | + | + | + | + |
| | | ± | ± | ± | + | + | + |
| B | | | | | + | + | + |
| | | | | | + | + | |
| | | | | | + | | |
| I | | | | | | + | |
| I and V | | | | | | + | + |
| V | | | | | | | + |

# General Conclusions on the Action of Colicins

Studies on tolerant mutants show that several genes can affect mediation of the colicin effect, presumably because several proteins are involved. The studies on LA639 suggest that each molecule of the Tol II protein becomes irreversibly associated with receptors soon after its synthesis. Since the Tol II protein is required for colicins A, C, E and K to act, it would seem that there are clusters of receptors of different types closely associated with Tol II protein and perhaps other proteins involved in mediating colicin killing.

It may well be that the proteins inferred from the existence of other tolerant mutants are also associated with colicin receptors and form a complex which plays an important role in regulating certain aspects of DNA, protein and energy metabolism. The pleiotropy of the CetC mutants and the conditional lethality of some tolerant mutants emphasise the basic nature of the functions controlled by the proteins involved. The increased sensitivity to detergents of some tolerant mutants suggests that the proteins involved are membrane associated.

We do not yet know the distribution of receptors over the cell surface, nor the nature of their association with the other proteins involved in the response to colicins. One possibility is that together these proteins form a network spread throughout the membrane, and that adsorption of a colicin molecule to a receptor causes some change of state to be propagated throughout the network (CHANGEUX and THIERY, 1967; NOMURA, 1967); this then affects the various biochemical targets, which, it is argued, are also associated with membrane. In the case of E3-CA38 action, this hypothesis was never very satisfactory as ribosomes probably do not ordinarily associate with membrane, and the observation that this colicin can act directly on ribosomes, and presumably enters the cell, renders it very improbable.

Perhaps a more likely possibility is that one or more complexes, comprising various receptors and other proteins, occur as discrete units in the membrane, and that their widespread influence in the cell is mediated by soluble repressors or activators of the enzymes involved in the ultimate lesion. This hypothesis does not require a complicated network able to transmit various sorts of information through the membrane. In the case of colicins of type E3, the receptor and associated proteins presumably effect the transfer of the colicin, or some active part of it, to the inside of the cell. The proteins encoded by the Tol genes may be involved in this transfer of colicin molecules across the cell envelope. Some Tol mutants affect colicins of types A, C, E1, E2 and K in addition to type E3, showing that there is some common step, again perhaps the transport of colicin, in the mediation of the effect of all these colicins. Other tolerant mutants — such as TolC, CetA, CetB and CetC may have a modified target as only one type of colicin action is affected.

Thus while studies of colicins have revealed the existence of a membrane associated control system, the details are far from understood. The fact that mutants tolerant to colicins E1 and K seem never to be tolerant to colicins Ia or Ib, producing a similar biochemical lesion, suggests that colicin I may not act through the complexes we have discussed but through yet another system.

## Bacteriocins Causing Membrane Damage

### Megacin A-216

This megacin has long been known to cause some membrane damage and recently it has become clear that this is the primary effect. IVANOVICS and ALFOLDI in 1955 observed that sensitive cells ceased to grow after addition of megacin and in fact the optical density and respiration rate decreased. In 1959 IVANOVICS, ALFOLDI and NAGY extended their studies using partially purified megacin and HOLLAND (1962) used very pure material. They found that the decrease in turbidity was due to leakage of cellular constituents, the cell wall generally remaining intact. Megacin A-216 also causes leakage from protoplasts (HOLLAND, 1962; IVANOVICS, ALFOLDI and NAGY, 1959; OZAKI et al., 1966). The effects are often long delayed, the delay increasing with decreasing concentration (HOLLAND, 1962; OZAKI et al., 1966). It was concluded that megacin caused a breakdown of the osmotic barrier without membrane dissolution.

Megacin A-216 has now been shown to be a phospholipase A (OZAKI et al., 1966) and in several details its action resembles that of plakin, an enzyme from blood platelets active on some Bacillus species, studied by the same group at Osaka (HIGASHI, SAITO, YANAGASE, YONEMASO and AMANO, 1966). Thus the activity spectrum can be extended by addition of bicarbonate and both show a delayed effect if diluted beyond a certain level. OZAKI and AMANO (1967) refer to unpublished data of HIGHASHI et al., suggesting that plakin is also a phospholipase A.

However, the specificity of megacin action does not seem to be explained as yet. *B. megaterium* itself is reported not to contain lecithin (KATES, 1964), the only substrate so far tested for the enzyme activity of megacin A. Thus the basis for the activity of megacin A against this species is not known, let alone why most other bacteria are resistant.

### Enterococcin 1-X14

The enterococcin of *Streptococcus faecalis* var *zymogenes* has been identified with the group D hemolysin of streptococci [BROCK and DAVIE, 1963; DAVIE and BROCK, 1966 (1)] although purified preparations or even preparations of high activity are not available. The identity is indicated by the absolute correlation in occurrence of the two properties in culture supernatants of streptococci and in the similar inactivation kinetics of the two properties.

The treatment of sensitive cells, either *Streptococcus faecalis* X46S or *Micrococcus lysodeikticus* leads to rapid loss of viability and ability to take up $^{14}C$ glycine (2 log drop in about 5 min) whereas X14, a producing strain, is unaffected and X46S-R1 (a resistant mutant of X46S) is affected much less. Similar effects were observed on spheroplasts, suggesting that cell wall was not involved in either sensitivity or resistance. The authors conclude that this bacteriocin has its effect on membrane and although all the data are compatible with this, it should be noted that there is as yet no direct evidence that the killing and loss of ability to take up glycine (the only metabolic function studied) do in fact result from such membrane damage. BASINGER and JACKSON (1968) found that sensitivity of bacteria to the hemolysin depended greatly on their physiological state, stationary phase cells being completely insensitive.

DAVIE and BROCK [1966 (1, 2)] have demonstrated an inhibitor of both hemolysin and bacteriocin activity which can be released from both X14 (the bacteriocin producer) and X46S-R1 (the resistant mutant) by suspension of the bacteria in acetate buffer. The inhibitor from both strains has been purified and found to be a teichoic acid with the following relative molar ratios: 1.00 mole phosphate, 1.05 mole glucose, approximately 1 mole ritibol and 0.62 mole of ester-linked D-alanine. The sensitive strain X46S produces a teichoic acid which is not inhibitory and lacks ester-linked alanine. Further the removal of ester-linked alanine destroys the inhibitory activity of the teichoic acids of X14 and X46S-R1. The properties of this inhibitor present perhaps the best evidence for the identity of the hemolysin and the bacteriocin and thus indirectly strongly support the hypothesis that the effect of the bacteriocin is on membrane, being identical to its action on red blood cells. The presence of the teichoic acid is presumably the basis of resistance of both X14 and X46S-R1 to bacteriocin and this is confirmed by their becoming sensitive if suspended in 0.2 M acetate buffer at pH 6, conditions which elute the teichoic acid.

## The Bacteriocin-like Effects of Phage Adsorption

Many bacteriophages inhibit host functions after adsorption. The T-even phages are perhaps the best-studied of the virulent phages, and their effects on host cell metabolism are apparently mediated in more than one way. Some hosts, such as

*E. coli* Co 270 or *Shigella dysenteriae* Sh (P2), can adsorb T-even phages and be killed by them but the phages do not propagate [BERTANI, 1953; FREDERICQ, 1952 (2)]. This lethal effect of adsorbed phage superficially resembles the action of colicin and in particular that of colicins of types E1, K, etc., in that all macromolecular syntheses are inhibited and permeases cease to function (LURIA, 1964). However, this phage effect is due to a complete breakdown of the permeability barrier (FIELDS, 1969) and hence differs from the known effects of any colicin. This breakdown of the permeability barrier of the cell is probably a prolongation of that which occurs briefly after infection of normal cells (PUCK and LEE, 1955). T-even phage ghosts (the protein coats free of DNA) inhibit bacterial synthesis and kill bacteria (HERRIOT and BARLOW, 1957; LEHMAN and HERRIOTT, 1958; LEVIN and BURTON, 1961). This effect may well resemble the abortive infection discussed above.

Under normal conditions, T4 infection inhibits host DNA synthesis and template function and the DNA is degraded. Host protein and nucleic acid synthesis is rapidly replaced by phage synthesis (BRENNER *et al.*, 1961; LEVIN and BURTON, 1961; NOMURA, HALL and SPIEGELMAN, 1960; TERZI, 1967; VOLKIN and ASTRACHAN, 1956). The degradation of host DNA is due to phage nucleases and is initiated by endonucleolytic cleavage. The other effects may follow from this enzymic damage to host DNA (WARREN and BOSE, 1968) and hence bear no resemblance at all to colicin action.

In the presence of CM, phage T4 still inhibits host RNA and DNA synthesis but the DNA is not degraded (NOMURA, MATSUBARA, OKAMOTO and FUJIMURA, 1962; NOAURA, WITTEN, MATEI and ECHOLS, 1966; TERZI, 1967). The inhibition in the presence of CM is markedly dependent on multiplicity and perhaps unrelated to the effects of T2 on Co 270 referred to above. In fact, because DNA, RNA and lipid synthesis are affected to different degrees, and because the cell can give rise to a burst after dilution of the CM, it seems unlikely that this inhibition is due to breakdown of permeability barriers, and it may resemble the effects of colicins on cells. The effect of T4 on *Shigella* resembles that on *E. coli* in the presence of CM and it would seem that here one can study an effect similar to that of colicins, without the complication of adding CM (TERZI, 1967). Host macromolecular syntheses are also repressed on infection by temperate phages such as λ (WAITE and FRY, 1964; SINSHEIMER, 1968; TERZI and LEVINTHAL, 1967) or P22 (FAVRE, AMATI and BEZDEK, 1968; SMITH and LEVINE, 1965). In the case of λ the phage DNA is involved since restricted phage have no effect, yet phage encoded protein seems not be involved since CM does not prevent the repression of RNA synthesis (TERZI and LEVINTHAL, 1967). The immunity of a λ lysogen, while preventing superinfecting phage from replication, does not prevent the effects of phage adsorption.

In the case of P22 and a related phage L, a brief effect on DNA synthesis is apparently due merely to phage adsorption and perhaps injection of DNA, and occurs even with a lysogenised host (FAVRE *et al.*, 1968). Further effects on total protein and DNA synthesis are due to phage encoded genes, as is the subsequent recovery of these functions necessary either for effective phage growth or lysogenisation (FAVRE *et al.*, 1968). A fascinating example of bacteriocin-like action due to a phage has been described by MARMUR and his colleagues. PBSX a defective phage of *B. subtilis* inhibits host DNA, RNA and protein synthesis within 5 min (OKAMOTO, MUDD, MANGAN, HUANG, SUBBAIAH and MARMUR, 1968). The DNA of this phage is largely of host origin and in any case is not injected so the effect on host metabolism is presumably a direct consequence of phage adsorption and comparable to the action of colicins. Studies of killing kinetics suggest that 10% of the adsorbed phages kill. The phage PBSX is produced by *B. subtilis* 168 on induction. This strain is not sensitive to the phage but is if cured of the lysogenic state. However, the parent strain does not have a typical immunity as the defective phage is not adsorbed to it.

## Summary and Conclusions

In this chapter we have seen that many bacteriocins with a known mode of action affect DNA metabolism, protein synthesis, or some aspect of energy metabolism.

Until recently colicins had been thought to act while bound to the surface receptors, and tolerant mutants to affect the chain of events leading to cell death. In the case of colicin E3-CA38 the colicin itself is now thought to enter the cell and directly interact with ribosomes which are irreversibly damaged, thus preventing protein synthesis. In this case tolerant mutants may be unable to effect the entry of colicin into the cell after adsorption to a receptor.

Another group of bacteriocins seem to have an enzyme-like action on membrane.

Bacteriophages can also have effects on cells in some cases very closely resembling the effect of bacteriocins.

Chapter 7

# The Bacteriocins in Nature

We have dealt at some length with what is known of the genetic characteristics of bacteriocinogeny and the lethal action of bacteriocins. Some of the phenomena we have discussed may appear bizarre, for instance the "lethal synthesis" discussed in Chap. 4 or the killing of a cell by a single protein molecule adsorbed to its exterior surface, discussed in Chap. 5. We can ask ourselves two general questions in this circumstance: what part do these phenomena play in the ecology of bacteria? and how did such a situation ever arise? Of course, there is yet much to be learnt about bacteriocins and it may seem early to speculate in fields where direct experimental evidence is effectively lacking. However, these are very interesting questions, and in this chapter we shall talk around them, knowing that we cannot hope to find a definitive answer. Because the colicins are so much better understood as a group than any other group of bacteriocins, we will largely confine our discussion to them. It is obvious that much will be equally applicable to many other groups of bacteriocins but sufficient is now known of two bacteriocins — megacin A-216 and enterococcin 1-X14 to suggest that they are very different and little of our discussion will be applicable to them.

## Evolutionary Origins of Bacteriocins and Their Role in Ecology

Ecological studies on the dynamics of bacterial population changes are rare, and none to my knowledge have adequately determined the role of bacteriocins in these changes. However, FLOREY et al. (1949) were able to cite several works which indicated antagonism between different bacteria in mammalian hosts. BRANCHE, YOUNG, ROBINET and MASSEY (1963), in a 6-month study of the intestinal *E. coli* of 5 people, found that those strains which were resident over long periods were usually colicinogenic whereas this was less common in the transient strains, and this is at least indicative that colicinogeny might play some ecological role *in vivo*. Various observations relating to this possible ecological role have been reviewed by VIEU (1964). Certainly the basic facts about bacteriocins suggest that they may be of considerable importance in natural populations of bacteria. They are extremely potent lethal agents, and the supernatant of a broth culture of a colicinogenic strain may be capable of killing many times the number of bacteria which were present in the producing culture, even perhaps several thousand times as many. Their activity on related strains may be of particular importance as producing and sensitive strains may be in direct competition, using the resources of the environment in much the same way.

The fact that the synthesis of a bacteriocin may be lethal would not detract much from the selective advantage of bacteriocinogeny as, if only a small proportion of the cells produce bacteriocin and die, a small advantage to the other bacteria in the same clone will offset this and give bacteriocinogeny a selective advantage. In most bacterial environments the members of a clone will tend to stay close together; this is particularly the case where there is a solid matrix, such as soil, and probably applies in a fluid environment with solid particles in it, such as gut contents. Under these conditions the production of a bacteriocin, even though it involves the death of some cells of the clone, may well reduce the ability of other bacteria of the same or related species to grow in the neighbourhood, and thereby provide the members of the bacteriocinogenic clone with increased access to the resources of the environment. Since colicinogeny normally also confers immunity to the same and similar colicins, the possession of a C-factor may confer a further selective advantage on the colicinogenic cell.

The possible biological advantages of bacteriocinogeny were, of course, readily apparent but we must now consider the biological reasons for sensitivity. This may not at first seem a sensible question, so let us compare bacteriocins with classical antibiotics such as penicillin. Penicillin is one of a group of antibiotics active against bacteria and interfering with their cell wall synthesis, probably by acting as an analogue of one of the cell wall components and inhibiting one enzymatic reaction in the assembly of the cell wall. The bacterial cell wall is unique in its structure; the biochemical pathways involved are not found in other organisms, including of course *Penicillium*, the fungus producing this antibiotic. Thus these other organisms are not affected by penicillin because of differences in the metabolism of fungi and bacteria. Other antibiotics probably also act as analogues of specific metabolites.

The action of colicins is quite different as they adsorb onto specific surface receptors which have no known function other than to act as receptors for bacteriocins and, perhaps, bacteriophages. These receptors do not seem to be essential for survival as they, or at least their ability to adsorb colicin, are readily lost by mutation without apparent adverse effects to the cell. The ability of a cell to die after a single protein molecule has adsorbed to the outside suggests that death results from a positive response on the part of the cell, which has a mechanism to respond to certain stimuli received at its surface. Thus we are forced to ask ourselves the function of these receptors whose presence on the surface renders the cell liable to be killed by a single bacteriocin molecule. An important clue here is that bacteriocins act only against strains of related species, suggesting that the specific interaction between bacteriocins and sensitive cells is a result of evolution within the species, perfected by natural selection for some advantage which accrues to both the cells partaking in this specific interaction [REEVES, 1965 (2)].

It is inconceivable that the described effects of bacteriocins could enable natural selection to maintain the existence of the receptor on the cell surface. This specificity must exist for some other function, and the one that comes to mind first is that it may be a recognition system, probably in connection with a fertility system of the type present in *E. coli* K12. On this hypothesis we suggest that the surface structure identified as a receptor for colicin activity is in fact a surface structure whose primary function is to mediate a suitable metabolic response during conjugation. Presumably these receptors are involved in the initial contact between bacteria of opposite mating type and, by their response to this initial contact, trigger off the events which lead to the successful transfer of chromosome from one bacterium to another and its incorporation during recombination. On this hypothesis, colicins are modified receptors which, on interacting with the complementary receptor elicit a response which is so exaggerated as to kill the cell.

Can we find any support for this hypothesis among the established data on bacteriocins? Perhaps surprisingly, several things fall into place if incorporated in this hypothesis. First, the association of several colicins with cell wall components is readily comprehensible, as this is the most likely location of any cell recognition factor. Second, derivation from a cell surface receptor might explain the lethal nature of bacteriocin synthesis, as if such surface components are released into the medium only after cell death, then it is clearly advantageous for a few cells to produce a lot of colicin and then die and so release it; the cause of cell death after colicin induction is not known. Third, the response of a sensitive cell to a single colicin molecule is explained, since during conjugation the area of contact is small and the cell must respond to this limited stimulus.

Does the observed response of sensitive cells to colicins appear plausible on this hypothesis? On this point it is difficult to be sensibly critical as we do not yet fully appreciate what is involved in bacterial mating, and hence, what components of metabolism one might expect to be vulnerable to any bactericidal agent acting on this particular system. (See CURTISS, CHARAMELLA, STALLIONS and MAYS, 1968, for a review of parental functions during conjugation.) A restriction of the synthesis of DNA or protein would appear to be a likely corollary of a cell's preoccupation with conjugation, involving the canalisation of its energy supply into either transfer or incorporation of a fragment of chromosome. An inhibition of energy metabolism is not so readily explained; it must also be admitted that there seems to be no good reason why the various aspects of a normal response to conjugation, those of protein synthesis and DNA synthesis for example, should each be independently elicited, and yet this would seem a necessary assumption to explain their being uniquely involved in the response to particular colicins.

Some observations on the physiology of conjugating bacteria also lend some support to our hypothesis. Thus during conjugation between strains of *E. coli* K12 the synthesis of $\beta$-galactosidase is reduced (RILEY, PARDEE, JACOB and MONOD, 1960) and indeed conjugation is sometimes lethal for the F$^-$ cell (CLOWES, 1963; GROSS, 1963).

A similar phenomenon of lethality by cell contact has been observed in *V. cholerae* (IYER and BHASKARAN, 1969; TAKEYA and SHIMODORI, 1969). It does seem that conjugation and bacteriocin action have something in common.

## Relationship of Bacteriocins to Bacteriophages

From the time that they were first discovered the similarities between bacteriocins and bacteriophages have been remarked upon, and it is sometimes suggested that they are related. The similarities occur essentially at three different levels. First we have the fact that both adsorb to specific receptors on the cell surface, and in some cases receptors of the same specificity may be used by both bacteriophages and colicins. We saw in Chap. 6 that some bacteriophages and defective bacteriophages produce effects which resemble those produced by bacteriocins, and so both may interact in similar ways with the receptor.

The second level at which bacteriocins and bacteriophages show similarity is in their inheritance. The DNA of C-factors and bacteriophages have both been classed as episomal (JACOB and WOLLMAN, 1958). However, the C-factors appear to be much more akin to F-factors than to bacteriophage DNA, which is why we have included both in the class of epichromosomes; the relationship between epichromosomes and bacteriophages is not clear although they do not appear to be closely related. Epichromosomes are able to replicate in phase with cell division and thus behave as typical chromosomes. In order to be amenable to study at present, an epichromosome must be capable of transfer to *E. coli* K12 or *S. typhimurium* LT2 and we do not know how many epichromosomes, like the main chromosome of *E. coli* K12, are incapable of promoting their own transfer to other bacteria, nor do we know if other fertile bacteria have chromosomes capable of promoting their own transfer. Thus epichromosomes in general may not differ in any essential way from the main chromosome.

Bacteriophages, on the other hand, are clearly parasitic in nature and their life cycle includes the complex mature phage with its protein coat and machinery for infection of a sensitive bacterium. The similarities between the two entities suggested by the properties which make them both episomes do not seem to imply a close evolutionary relationship. Perhaps the prophages which do not have a specific locus on the bacterial chromosome (IKEDA and TOMIZAWA, 1968; JACOB and WOLLMAN, 1961; TAYLOR, 1963) may bridge the gap between epichromosomes and bacteriophages, and NOVICK (1969) in a review of extrachromosomal inheritance places members of the two groups in a continuous spectrum.

The third level of similarity between epichromosomes and bacteriophages is that synthesis of some bacteriocins and of some bacteriophages is inducible.

At present we do not understand the mechanism whereby either bacteriophage or bacteriocins are induced, although this was discussed in Chap. 4. Superficially at least the two phenomena appear very similar, but only when they are understood will we know whether this similarity indicates an evolutionary relationship between bacteriophages and bacteriocins.

The possible interrelations between bacteriophages and bacteriocins are complicated by the possibility that phage DNA and epichromosomes may have undergone recombination giving recombinant factors. Thus FREDERICQ [1963 (3)] has suggested that some phages may have incorporated genes for colicin production into their genome, and the colicin itself into their tail structure, to account for the colicin-like effects of some phages on sensitive cells.

## Paramecium and Paramecin

Anyone familiar with the works of SONNEBORN, BEALE and others on *Paramecium* will have constantly noticed similarities between the observations on bacteriocins described in this book, and the work on *Paramecium* so well reviewed by BEALE (1954). For those who are not familiar with this work we can summarise it briefly. Some strains of *Paramecium aurelia* carry genetic entities in their cytoplasm called kappa particles, which cause the production of a protein called paramecin, lethal for some other strains of *Paramecium*. The remarkable resemblances lie in the fact that the genetic entity can be transferred from one *Paramecium* to another and confers immunity to paramecin, and that one molecule of paramecin can kill a sensitive *Paramecium*. One might, therefore, apply the same arguments we have used above to suggest that we are dealing with a modified response to conjugation. *Paramecium* has of course a conjugation system, and it is interesting that the response of a sensitive cell to paramecin is said to superficially resemble the early stages of conjugation, and that after a time the meganucleus starts to disintegrate, again resembling the sequence of events during conjugation (BEALE, 1954). A further remarkable similarity between paramecin and bacteriocins is that in some cases the sensitivity of a *Paramecium* to paramecin is correlated very closely with the presence of a particular surface antigen on the surface of the sensitive cell, indicating that the paramecin may adsorb to the cell surface. The hypothesis that we introduced to explain the existence of bacteriocins thus seems to be equally capable of explaining the existence of paramecin. In fact, the closer we look the more remarkable is the ease with which one can explain the phenomena reported for paramecin with our hypothesis. Thus during conjugation, Paramecia are protected from the killing action of paramecin, and one may speculate that, if the effect of paramecin on a cell constitutes an incomplete conjugation stimulus, then during conjugation the complete stimulus would be "dominant" over any such incomplete stimulus.

Mu, another cytoplasmic particle of *Paramecium*, resembles kappa in its inheritance but causes the killing of mates during conjugation, rather than the release of a killing principle like paramecin. This phenomenon further confirms our suggestion that such killing activity involves modified conjugation responses. In this case we suppose that the specific agent retains its primitive location on the cell surface, while yet having been modified to evoke a lethal response from its mate. We should note, however, that the kappa and mu particles appear to be symbiotic bacteria (BEALE and JURAND, 1960, 1966) and do not closely resemble the C-factors in structure. We should not perhaps labour the similarity between bacteria and *Paramecium* too long, but let us at least note that these two groups — the bacteria and the ciliates, are the only two in the whole plant and animal kingdoms in which after conjugation at the cellular level both exconjugates remain independent and viable, although one or both now carry genetic material received from the other. In the case of ciliates, the genetic material transferred always consists of a complete nucleus, and we then have a typical fusion of two nuclei. Nonetheless, the situation in ciliates differs markedly from the usual eucaryote fertilization where two cells fuse completely, and, in its partial transfer of genetic material between two cells which eventually go their own way, closely resembles bacterial conjugation. The significance of this particular similarity for our hypothesis is not evident, but the number of analogies between bacteria and ciliates seems remarkable and perhaps worthy of comment even if not yet explicable.

## The Killer Characteristic in Yeast

The killer characteristic of yeast (SOMERS and BEVAN, 1969; WOODS and BEVAN, 1968) is due to a protein made by killer strains; the killer characteristic is dependent on the presence of cytoplasmic particles whose maintenance is in turn dependent on a dominant nuclear gene. The producing strains are immune to the protein, which kills other strains of yeast.

The situation resembles that described for bacteria carrying C-factors and even more closely that for *Paramecium* carrying kappa particles, in which the maintenance of the particles is also dependent on a dominant nuclear gene (see BEALE, 1954, for review). It would seem that phenomena resembling some of those we have discussed for bacteriocins may occur in organisms widely divergent in evolutionary history.

# A Catalogue of Bacteriocins

In chap. 1 we referred briefly to bacteriocins other than colicins and in the intervening chapters we have discussed some of these other bacteriocins in some detail. In this appendix we shall look at each of the groups of bacteriocins currently known to see how they compare.

## Bacteriocins of the Enterobacteriaceae

In the early work on colicins, several genera were used and although the bacteriocins of *Citrobacter* and *Shigella* seemed even then to be distinctive (FREDERICQ, 1948) they were nonetheless included as colicins. When HAMON and PERON and others came to study genera more distantly related to *Escherichia* they identified several new families of bacteriocins, which while showing overall similarities to the colicins seemed to have a different specificity, each being active primarily on strains of the same genus or group of genera. Some of these have only been briefly reported on while a considerable amount of work has been done on others.

### Alveicins
#### (named after *Hafnia alvei*)

This family was not defined in the usual manner. Ninety-eight *Hafnia* strains were tested against 2 strains of *Hafnia* and 4 common indicator strains for colicins [HAMON and PERON, 1963 (1)]. Thirty percent were active against some of the colicin indicators but none against any of the *Hafnia*. The alveicins were distinguished from the colicins because they act on only some of the 4 colicin indicators, whereas colicins generally kill all 4. The lack of activity against *Hafnia* may perhaps have been due to the small number of strains tried as indicators. The status of the alveicins seems uncertain, but until the matter has been studied further it seems preferable to follow HAMON and PERON and separate them from colicins.

### Arizonacins
#### (named after *Paracolobactrum arizonae*)

Of 200 strains of *P. arizonae* studied, 15% produced a bacteriocin active against one or more of the 3 *E. coli*, 1 *Shigella* and 25 *P. arizonae* strains tested [HAMON and PERON, 1963 (1)]. A few of these acted on only one of the *Paracolobactrum* strains; the majority acted on several of them and in addition one, two or three of the other strains (the three *E. coli* and the *Shigella* were usual colicin indicators).

## Caratovoricins

### (named after *Erwinia caratovora*)

Of 9 strains of *Erwinia* tested [HAMON and PERON, 1961 (2)], 7 produced a bacteriocin and their activity was tested against the 9 *Erwinia* strains and several *Pseudomonas, Serratia, E. coli* and *Shigella* strains. Only 2 of the 7 had identical spectra, and of the 6 types, 3 did not include any *Erwinia* in their spectrum at all. There was, in general, considerable activity on *Pseudomonas, E. coli* and *Serratia*, with 11 out of 63 possible interactions for *Erwinia* and 18 out of 77 possible interactions with the other genera being positive. Once again none were active on all of the usual colicin indicator strains used.

## Cloacins

### (named after *Enterobacter cloacae*)

Of 29 strains of *Enterobacter* tested [HAMON and PERON, 1963 (4)], 27% produced bacteriocins when tested against the same 29 strains and 2 *E. coli* strains (K12 and B). All were active against one or more of the *Enterobacter* strains used and some were also active against one or both of the *E. coli* strains. In this study the effect of ultraviolet induction on the producing strains was examined and the activity spectra were changed (rather than extended), suggesting that a different cloacin might be produced on induction. PAPAVASSILIOU [1960 (2) and personal communication] found some strains of this group to produce a bacteriocin active on *E. coli* and giving cross-resistance with colicin I. STOUTHAMER's group have studied cloacin DF13 in some detail as related in the appropriate chapters. It is active on many *Klebsiella* strains (STOUTHAMER and TIEZE, 1966; DE GRAAF and STOUTHAMER, 1971).

## Marcescins

### (named after *Serratia marcescens*)

Several species of *Serratia* can produce marcescins active both on other *Serratia* strains and on *E. coli*, and 85% to 100% of strains are bacteriocinogenic [HAMON and PERON, 1961 (1); MANDEL and MOHN, 1962].

HAMON and PERON [1966 (4)] studied trypsin inactivation of marcescins in culture supernatants. The activity against *E. coli* strains was in general destroyed whereas that for *Serratia* strains was not, demonstrating the existence of at least two groups of marcescins, one trypsin-sensitive and active on *E. coli* and perhaps on *Serratia*, the other resistant and active only on strains of *Serratia*. Many strains seem to produce at least one of each type.

The marcescins of fraction 1 (the trypsin-sensitive group) are heterogeneous as shown by neutralisation studies using specific antisera [HAMON and PERON, 1965 (1), 1966 (3)]. They also show some relationship to colicins and in particular colicins of type E, on the basis of cross-resistance and immunity.

In contrast to the studies of HAMON and PERON we have the claim made by MANDEL and MOHN (1962) that the marcescins active on *E. coli* show cross-resistance with colicin K, and this has been confirmed by PRINSLOO, MARE and COETZEE (1965), using mutants of *E. coli* φ. They further found that all marcescins active on *E. coli*, *Hafnia* or *Aerobacter* were not mobile on agar electrophoresis, while almost all of those active on *Serratia* migrated to the cathode and all but one with the same mobility. The colicin K used did not migrate like fraction 1 marcescins. PRINSLOO (1966), in a further study of the two groups of marcescins, referred to those which HAMON called fraction I as group B, and fraction II as group A. He found that the group B marcescins were active on *Hafnia* and *Aerobacter* as well as *Escherichia*, that antisera to one would neutralize all, and that resistant mutants are resistant to all; they consider all 54 to be identical, with one exception. All the group B marcescins were heat- and trypsin-labile in agreement with HAMON and PERON's observations. The group A marcescins showed seven distinct activity spectra, using both wild type indicators and resistant mutants. The existence of resistant mutants which show specificity is analogous to the colicin situation. Several other gram-negative and gram-positive genera were included in the tests but none was sensitive to either group A or group B.

One class of marcescins then are active on *E. coli* and related organisms, and are considered by one group to be related to colicin E and heterogeneous, and by another to colicin K and homogeneous; a second class of marcescins are active on *Serratia* and show heterogeneity resembling that of colicin types. The existence of two such classes is unusual but the marcescins seem clearly to be bacteriocins.

## Pneumocins

### (named after *Klebsiella pneumoniae*)

*Klebsiella* strains were included in some of the series tested for colicin production [FREDERICQ, 1948; LINTON, 1960; PAPAVASSILIOU, 1960 (2)], and some did act on the classical colicin indicator strains. Detailed surveys of the various species in this genus have shown that about 35% produce bacteriocins, active either on other *Klebsiella* strains or sometimes on *E. coli* and *Shigella* strains [DURLAKOWA, MARESZ-BABCZYSZYN, PRZONDO-HESSEK, LUSAR and MROZ-KURPIELA, 1964 (1, 2); HAMON and PERON, 1963 (1); MARESZ, DURLAKOWA, HAMON and PERON, 1966; MARESZ, HAMON and PERON, 1968; STOUTHAMER and TIEZE, 1966].

In some of these studies the aerocins of *Klebsiella aeruginosa* are separated from the pneumocins, but at present the distinction does not seem justified.

It is possible to recognize many different groups of pneumocins on the basis of their activity spectra on other *Klebsiella* strains [DURLAKOWA *et al.*, 1964 (1, 2); HAMON and PERON, 1963 (1); MARESZ *et al.*, 1968; MARESZ-BABCZYSZYN, DURLAKOWA, MROZ-KURPIELA, LACHOWICZ and SLOPEK, 1967; SLOPEK and MARESZ-BABCZYSZYN, 1967; STOUTHAMER and TIEZE, 1966] and it is also possible to isolate resistant mutants which show specificity, allowing several pneumocin groups to be recognized (DE GRAAF and STOUTHAMER, 1971; MARESZ *et al.*, 1966; MARESZ-BABCZYSZYN, DURLAKOWA, LACHOWICZ and HAMON, 1967; STOUTHAMER and TIEZE, 1966). However, many of the mutants are resistant to all or most pneumocins. The various species of the genus differ in the proportion which produce or are sensitive to pneumocins and in the types of pneumocins produced. Some at least are determined by epichromosomes as was discussed in Chap. 3. Some can be neutralised by antisera for colicins of type I (MARESZ et al., 1966).

DE GRAAF and STOUTHAMER (1970, 1971) have isolated tolerant mutants of *Klebsiella pneumoniae* which, although no longer sensitive, can still adsorb a particular pneumocin or cloacin. Many of the mutants are tolerant to the three pneumocins tested, and also to colicin A-CA31. Many mutants are tolerant to some pneumocins and resistant to others.

The pneumocins active on *E. coli* strains are probably distinct from those acting on *Klebsiella* strains [HAMON and PERON, 1967 (2); MARESZ et al., 1968] and can be subdivided on the basis of their activity spectra and the cross-resistance of mutants. Some show cross-resistance with colicins of type B (MARESZ et al., 1968) and others with colicins of types E, I or K [HAMON et al., 1970; PAPAVASSILIOU, 1960 (2)]. Some strains produce more than one pneumocin and these can be distinguished by use of differential inactivation by heat, antisera, or trypsin (MARESZ et al., 1968).

## Miracins, Morganocins and Provicins

### (named after *Proteus mirabilis, P. Morgani* and *Providencia*)

The *Proteus* group, including the very closely related *Providencia*, has recently been examined for bacteriocin production. In the first study CRADDOCK-WATSON (1965), using 18 selected strains as indicators, found that 139 (60%) of 229 strains of *P. mirabilis* produce a bacteriocin. The activity spectra on the 18 *P. mirabilis* made it possible to identify 26 different spectra, but 90 of the 139 fell into 3 of these types. The activity of these bacteriocins against other *Proteus* strains was not tested, but the 3 major types were not active against *E. coli* K12 or some *S. sonnei* strains. It is of interest that one *P. rettgeri* out of 12 studied produced a bacteriocin and it was identical in spectrum to one of the 3 major types.

COETZEE (1967) used 94 *P. morgani*, 32 *P. rettgeri*, 28 *P. hausseri* and 37 *Providencia* strains. In the initial search he tested each of the above strains only intraspecifically and then tested each bacteriocin found against all the other strains. Thus any bacteriocin active only outside its species would not have been detected. Five of the 37 *Providencia* strains produced a bacteriocin of which only one acted outside the *Providencia* strains and that only on one strain of *P. rettgeri*. All 5 had distinct activity spectra. Twelve of the *P. morgani* srtains were bacteriocinogenic but none of the morganocins acted on the other *Proteus* species or on *Providencia*. All 12 had distinctive activity spectra within *P. morgani*.

Thus there seem to be at least three distinct bacteriocin families within the *Proteus-Providencia* group.

## Colicins

Having looked at the other bacteriocin types, we must have a closer look than was done in Chap. 1 at some of the details of colicin classification, before considering the interrelationships of all the bacteriocin families of the Enterobacteriaceae.

We have seen earlier that FREDERICQ recognized 12 different colicin types on the basis of the specificity of resistant mutants: A, B, C, D, E, G, H, I, K, V, S1 and S4. All of these are still extant and type strains have been designated [FREDERICQ, 1965

(1)]. Several new types, L, M, N, O, P, Q and X, have been added although the status of some is not certain.

FREDERICQ [1953 (1)] discovered one strain producing an unknown colicin and called it type L. No other examlpes of this colicin seem to have been reported. In 1962, HAUDUROY and PAPAVASSILIOU reported another new colicin, which they also called type L, and again no other examples have been reported. The two colicins with this designation are not known to be related. HAUDUROY and PAPAVASSILIOU studied the cross-resistance of mutants resistant to their colicin L, and concluded that it was related to the E group of colicins.

Colicin M, discovered by FREDERICQ [1951 (1, 2, 4, 5)], is produced by mutants of *E. coli* CA7 resistant to phages T1 and T5, whereas the parent strain CA7 produces only colicin V. Since colicin M was found to show cross-resistance with these two phages, it seems that strain CA7 probably adsorbs all the colicin M it produces whereas the phage-resistant mutants, lacking the common receptor, release the colicin into the medium. This colicin may be the same as that described by BORDET (1947) as produced by CA7. The situation is not simple, however, and colicin C resistant mutants of CA7 also produce colicin M and show cross-resistance to phage T1. One might conclude that colicins C and M adsorb to the one receptor, but in fact the situation is probably yet to be resolved. Certainly colicins C and M differ in some respects, such as activity on *E. coli* B and on CA7 itself. Colicin M is also produced by other strains in conjunction with colicin B (FREDERICQ and SMARDA, 1970).

Colicin N was described by HAMON and PERON [1964 (1)] and KASATIYIA and HAMON (1965) as produced by 5 strains of *E. coli* of serotype 0119:B14. It was thermostable, but active on mutants resistant to the other known thermostable colicins, A, D, K and V, and hence placed in a separate type.

Colicin P, also thermostable, was described in the same paper as colicin N and was produced by one strain of *E. coli* of serotype 055:B5. Mutants of *E. coli* K12 showed no cross-resistance between P and A, K or V but all mutants which resisted P also resisted D and N. They distinguished it from these two colicins by its restricted activity spectrum, but in view of the variability of colicins of some other types, it seems possible that colicins P, D and N might all belong to one type.

The designation P, like L and X, has been used twice, having previously been used by FREDERICQ [1953 (1)] to refer to the colicins of a group of 50 strains. These colicins were perhaps related to colicins G and H and do not seem to have been studied further.

Colicin X was described by PAPAVASSILIOU [1961 (1)] as produced by one strain of *E. coli*. It showed no cross-resistance with colicins A, B, C, D, E, I, K, L or V, the other colicins not being tested. It was very active against *E. coli*, 92.5% of strains being sensitive, a considerably higher percentage than for most other colicins. MIYAMI, OZAKI and AMANO (1961) also described a colicin X, produced by *E. coli* K235 in addition to colicin K.

Colicin O does not seem to have been fully described as yet although it is sometimes referred to [e.g. HAMON and PERON, 1964 (4)]. Colicin Q (SMARDA and OBDRZALEK, 1966) shows complete cross-resistance with colicin V and so if a classification on this basis is used it is a variant of colicin V. Other colicins which have been described include that of *E. coli* 15, since shown to be a defective bacteriophage (see Chap. 2) with unusual properties worthy of brief mention. It acts only on the producing strain, *E. coli* 15, and resistant mutants simultaneously acquire sensitivity to 6 of the 7 T-phages (MUKAI, 1960). If phage-resistant revertants are selected from such a mutant, then some of them are again sensitive to the "colicin" and this correlates with simultaneous resistance to T1, T3, T4, T5 and T7.

The colicin produced by *E. coli* B (COCITO and VANDERMEULEN-COCITO, 1958) seems to be atypical, being produced in barely detectable amounts.

MESROBEANU, MESROBEANU, CROITORESCO and MITRICA (1964) tested some neurotoxins prepared from *S. typhimurium, Shigella, Escherichia, Proteus, Pseudomonas* and *Serratia* and found them all to be very active against the producing strains, all other gram-negative strains tested and also some gram-positives. Although their activity spectra are highly atypical for bacteriocins of gram-negative bacteria, it is not clear why these substances have not shown up in the many surveys for bacteriocins. They are lipoprotein and were identified serologically as part of the O antigen complex.

We have referred to some of the colicin types many times already, but there are some points which have not been mentioned and are worthy of discussion here.

Colicin D has, with colicin X, the highest frequency of activity on *E. coli* and *Shigella* strains, acting on 94 and 92% respectively and even on 88% of *Salmonella* strains. Specifically resistant mutants are not common. Colicins of this type are apparently not all identical [FREDERICQ, 1948 and 1953 (1)].

Colicins of type E, as we saw in Chap. 1, have been the subject of very intensive study. We considered there the subdivision into subtypes E1, E2, E3 and E4 and also subdivision on the basis of mutational tolerance, and these topics were considered again in Chap. 4 and 6. It is clear that colicins of type E are highly variable and yet more variation will be discussed in the ensuing section on activity spectra, but most authors nonetheless consider all these colicins to adsorb to the one receptor. However HAMON and PERON [1966 (1)] consider colicin K30-E1 to adsorb to a different receptor than their colicins representing E2 and E3. Their arguments rests on a series of mutants which either resist colicin E1 and not the others or vice versa. They state that these mutants are receptor mutants as they no longer adsorb the colicins they resist, and HILL and HOLLAND (1967) also report such mutants. However, until data are available for other colicins of each type, it seems better not to separate colicins E2 and E3 under the new type F.

The C-factors determining colicins of type E1 are divided into two types, E1a and E1b, depending on their ability to undergo epidemic spread (LEWIS and STOCKER, 1965). However the colicins have not been shown to be different.

Colicins G and H are very similar in their inhibition zones and specifically resistant mutants are not easily obtained. Colicin H has been reported to be the only colicin active on *Proteus* (FREDERICQ, 1948), although other reports are conflicting (SMARDA and SCHUHMANN, 1966), while colicin G is reported to be the only colicin active on *Pasteurella* (SMITH and BURROWS, 1962). SMARDA [1966 (1)] found a specific inhibitor for these two colicins which was produced by a strain of *Proteus mirabilis*. It is conceivable that the inhibitor is in fact free receptor substance, but in any case it once again demonstrates the similarity of these two colicins, which seem in several ways to be atypical.

Colicin I has been discussed already in Chap. 1 because it can be subdivided on the basis of immunity. Colicins of this type have been studied frequently, particularly in genetical work, but the fact that high titre preparations are difficult to obtain has limited chemical and even mode of action studies. It gives frequent cross-resistance with colicin V but produces a much smaller inhibition zone [PAPAVASSI-LIOU, 1960 (3)].

Colicin K is one of the better known because K-K235 has been the subject of the extensive chemical investigation of GOEBEL and his colleagues. Two colicins put into this type differed from the majority and if really of type K indicate that this colicin type is also heterogeneous [FREDERICQ, 1953 (1)].

Colicin V has the largest inhibition zone of the colicins apart from a colicin produced by *Salmonella* (ATKINSON, 1967), and is presumably of correspondingly low molecular weight. The problem of its dialysability was discussed in Chap. 2. This colicin has been used in various genetic studies but, as for colicin I, with which it gives frequent cross-resistance, it is difficult to obtain in sufficiently high titre for many studies although HUTTON and GOEBEL were able to purify the colicin V-K357.

## Colicins — Activity Spectra

We saw earlier that FREDERICQ (1948) in his study of the action of 88 colicinogenic strains on 316 strains of various genera of the family, could recognise 17 distinct activity spectra but that he later decided that 6 of these groups (E, F, J, S2, S3 and S5) should be considered as one colicin type on the basis of resistant mutant specificity. However, the variation in activity spectra of colicins of type E has recently received further attention.

HAMON and PERON [1966 (2, 4, 7)] examined the activity spectra of various colicins produced by *Shigella*, *Salmonella* and *Citrobacter*. They showed that most *Shigella* colicins are of type E, on the basis of cross-resistance of mutants of *E. coli* strains. They used their mutants specifically resistant to E1 or to E2 and E3 and were able to classify the *Shigella* colicins into one of these two subgroups. Likewise, the *Salmonella* colicins included several which on the same grounds were considered to be of type E. However, the activity spectra of these colicins showed considerable variation (12 different spectra for the *alkalescens* — *dispar* group of *Shigella* alone) and also differed from the colicin E of *E. coli*. For instance, while all colicins of *E. coli* are said to act on *E. coli* K12, B and 36 and on *Shigella* Y6R, many of these colicins acted on only one, two or three of them and some on none. Likewise, the colicins B and I of *Salmonella* strains and colicins of type A of *Citrobacter* show variations in activity spectra and on this basis HAMON and PERON [1966 (2, 4)] would separate both groups from the colicins, although there is considerable overlap of activity spectra.

The variation in activity spectra of *S. sonnei* colicins has been used as a basis for typing strains of this species, which account for a high proportion of outbreaks of *Shigella* dysentery (ABBOTT and SHANNON, 1958; ABBOTT and GRAHAM, 1961). Sixteen distinct types are recognised in this scheme which uses 15 indicator strains, 13 of which are also *S. sonnei* strains and about half of which are resistant mutants of other indicator strains. This typing scheme has been widely used and commonly 50—70% are typable (ABBOT and SHANNON, 1958; ABBOTT and GRAHAM, 1961; BARROW and ELLIS, 1962, 1967; COOK and DAINES, 1964; GILLIES, 1964, 1965; HART, 1965; KAMEDA, HARADA and MATSUYAMA, 1968; LASZLO and KEREKES, 1969; MOTOKI,

USHIO and YOSHINO, 1968; MULCZYK, KRUKOWSKA and SLOPEK, 1967; SLOPEK, MULCZYK and KRUKOWSKA, 1967; STEPANKOVSKAYA and BRUTMAN, 1968; TOKIWA, SAKAMOTO and KAJIKURI, 1967). The various types occur in different proportions in different districts and from year to year, and the scheme has proved useful in epidemiology. Some of the papers describe new colicinotypes in additions to the 16 of the original papers.

NAITO, KONO, FUJISE, YAKUSHIJI and AOKI (1966) have independently developed a typing scheme for *S. sonnei*. The system was developed using 76 strains of which 54 were colicinogenic, inhibiting one or more of the same 76 strains. From these 76 strains 8 indicators were selected which permitted classification of the 76 strains into types A, B, C, D and E on the basis of their activity spectra, A being the non-colicinogenic strains. Each of the Groups A, B and C (which include the vast majority of strains), can be divided into 2 subgroups on the basis of their sensitivity to the colicins produced by the same 8 indicator strains (all of which are colicinogenic), thus making 8 groups in all, A1, A2, B1, B2, C1, C2, D and E. Of 1,010 other strains tested in this scheme, all but 39 were classifiable although 258 were in types A1 and A2 which are non-colicinogenic in these tests. AOKI, NAITO, FUJISE, IKEDA, MIURA and YAKUSHIJI (1967) compared the two typing systems for *Shigella* using 462 strains. There were considerable differences between the two systems although some types coincided exactly or nearly exactly with types in the other system. The heterogeneity of *Shigella* colicins is obviously greater than indicated by either typing system alone. It is not yet known how most of these types fit into the FREDERICQ classification, although most are likely to be variants of colicin E. Two are reported to be of types Ia and Ib (STOCKER, 1966).

The variation in activity spectra of colicins is not random, there commonly being a tendency for bacteriocins to act more frequently on closely related strains. Thus within the colicin group FREDERICQ (1948) found colicins of *Shigella* to act on *Shigella* strains more frequently than do colicins of *E. coli*, and HAMON and PERON [1966 (2)] find even within colicins of types E or I that those produced by *Salmonella* are more frequently active on *Salmonella* strains than those of *E. coli*.

### The Colicinogenic Bacteria

Despite the enormous variation in activity spectra within some colicin types, such as type E, classification on the basis of cross-resistance seems to be straightforward and most colicins fall into relatively few types. This limited classification has been in use for some time now, and from the several surveys of colicinogenic bacteria it is evident that various genera and species differ in the colicin types produced. About 25—40% of *E. coli* strains produce colicins [e.g., FREDERICQ, 1948, 1963 (1); VOSTI, 1968], which include most of the types so far described, including types A and K, which were not produced by any *E. coli* strains originally studied by FREDERICQ. FREDERICQ's early work (1948) showed that the colicinogenic property was not distributed randomly among the different groups of *E. coli* and later work has suggested some difference between pathogenic and non-pathogenic varieties of *E. coli*, and also between *E. coli* strains derived from various sources [FREDERICQ, 1950 (3); FREDERICQ and BETZ-BARREAU, 1950; FREDERICQ and JOIRIS, 1950; FRE-

DERICQ, BETZ-BARREAU and NICOLLE, 1956; GOUPILLE, HAMON and VIEU, 1960; HAMON, 1958 (2), 1959, 1961; KASATYIA and HAMON, 1965; KUDLAI, LIKHODED and GOLUBEVA, 1964; PAPAVASSILIOU, 1960 (2)]. Some of the more notable correlations are worthy of mention. For instance, of colicinogenic strains of serotype 055:B5 from cases of infantile gastroenteritis in Russia, 96% produce colicin G (KUDLAI et al., 1964). Other serotypes isolated from similar cases are reported to produce colicins of types B, D, E, I and G. [FREDERICQ et al., 1956; HAMON, 1959; KUDLAI et al., 1964; PAPAVASSILIOU, 1961 (2)]. About 300 strains were involved in these studies and the absence of other colicin types is significant. A second example concerns the frequent occurrence, in stools of people suffering from Salmonella paratyphi B, of E. coli strains producing colicin B, which is about the only colicin active on this Salmonella species. [FREDERICQ, JOIRIS, BETZ-BARREAU and GRATIA, 1949; FREDERICQ, 1950 (3)].

The colicins produced by repeated isolates from an epidemic are usually the same and colicinogenotyping can be used for epidemiological studies of E. coli (HAMON, 1959, 1961; McGEACHIE, 1965).

Citrobacter, unlike E. coli, can produce very few colicins. FREDERICQ (1948) and HAMON and PERON [1966 (4)] found only colicin A to be produced, although PAPAVASSILIOU [1960 (2)] also reported production of colicins E, I and K. It is a peculiar fact that Citrobacter strains producing colicin A have been found almost exclusively among strains isolated from the urine of people infected with Salmonella typhi and conversely most typhoid patients (15 out of 22 in the series studied) secrete such organisms. These particular Citrobacter strains also differ from typical strains in certain biochemical properties, and we thus have a very high correlation between colicin A production and these biochemical characteristics [FREDERICQ and BETZ-BARREAU, 1948 (1, 2)].

Shigella strains produce only types E, B, D, I and K, but there are considerable differences between the various Shigella species (CEFALU and BAVASTRELLI, 1959; FREDERICQ, 1948; HAMON and PERON, 1966 (7); HUET, PAPAVASSILIOU and BONNE-FOUS, 1961; PAPAVASSILIOU, VANDERPITTE, GATTI, and DE MOOR, 1964; STAVSKII, KNYSH and FOMICHEV, 1968]. It should be remembered that some Shigella colicins do not act on the normal colicin indicators and so they are at present untypable and may or may not be variants of the types already reported for this genus. S. dysenteriae has not been found to produce bacteriocins while S. sonnei is frequently colicino-genic (50—100% of strains) and is usually reported to produce colicins of type E only, as also for S. alkalescens and S. dispar. However, HAMON and PERON [1966 (7)] report that colicinogenic strains of Shigella species usually produce a small amount of colicin I as well as colicin E and this is well established for S. sonnei strain P9, which is in common use. The proportion of colicinogenic strains of S. flexneri varies from 3% to 30% in different reports [CEFALU and BAVASTRELLI, 1959; HAMON and PERON, 1966 (7); HUET et al., 1961; MULCZYK, SLOPEK and MARCINOWSKA, 1967; STAVSKII et al., 1968; VASSILIADIS, PATERAKI and POLITI, 1964]. HAMON and PERON found them to make E with some concomitant I production, but CEFALU and BAVA-STRELLI (1959) and HUET et al. (1961), while finding fewer colicinogenic strains, did not find the same overwhelming preponderance of E and report I and K. There are also many untypable colicins present, some of which may perhaps be variants of the known types. S. boydii produces E, I, B and D (HUET et al., 1961). However, although

*Shigella* produces only a limited number of colicin types, the strains producing colicins of a given type can, as we have seen above, have very variable activity spectra.

*Salmonella* strains are restricted in the colicins they produce to types E, B, I and K. About 10% of *S. typhimurium* are colicinogenic and usually produce colicin I or rarely colicins K or E (either E1 or E2) [FREDERICQ, 1952 (1); HAMON and PERON, 1964 (2), 1966 (2); LEWIS and STOCKER, 1965; OZEKI, 1960; OZEKI *et al.*, 1962; PAPAVASSILIOU, 1960 (1); PAPAVASSILIOU and SAMARAKI-LYBEROPOULOU, 1957; VASSILIADIS, PAPAVASSILIOU, GLAUDOT and SARTIAUX, 1960].

Other species of *Salmonella* are occasionally colicinogenic and these usually produce type I or less commonly type B. *S. typhi* is exceptional in that lysotypes 36 and 40 are almost always (98%) colicinogenic and produce colicin B (OZEKI, 1960; HAMON and NICOLLE, 1962). Such a high correlation of a colicinogenic property with other characteristics is very unusual. Other lysotypes of *S. typhi* are only rarely colicinogenic; the colicin produced is usually of type B, although one strain produced a colicin of type E (NICOLLE and PRUNET, 1964).

### Sensitivity of Different Genera to Colicins

Most strains of *E. coli*, *Shigella* or *Salmonella* are sensitive to one or more colicins (BELAYA *et al.*, 1969; CEFALU and BAVASTRELLI, 1959; FILICHKIN, 1968). However, whereas the activity spectra of the 17 type colicins all include some *E. coli* and *Shigella* strains, many do not include any *Salmonella* strains. Even those colicins which can act on *Salmonella* usually act on relatively few strains (FREDERICQ, 1948; PAPAVASSILIOU, 1965).

The species of *Shigella* vary in the proportion of strains sensitive to particular colicins (CEFALU and BAVASTRELLI, 1959; FREDERICQ, 1948; FILICHKIN, 1968; PAPAVASSILIOU *et al.*, 1964). In colicin sensitivity, as for colicins produced, there are often differences in the strain collections used by different authors. Thus FREDERICQ (1948) found all of 7 *S. sonnei* to be sensitive to most colicins, including A and S5, whereas PAPAVASSILIOU *et al.* (1964) find 0% and 4.9% respectively to be sensitive to these two.

The strain variation in sensitivity to colicins is probably not useful for typing *S. sonnei* strains, due to the variability amongst isolates from a given focus (ABBOTT and SHANNON, 1958; NAITO *et al.*, 1966), although there are many colicinotypes for most *Shigella* species (CEFALU and BAVASTRELLI, 1959). However, CEFALU and BAVASTRELLI (1959), CHAKRABARTY (1964), PAPAVASSILIOU and HUET (1962), VASSILIADIS (1964) and VASSILIADIS *et al.* (1964) all consider colicinotyping to be potentially useful for *S. flexneri*, where colicinogenotyping is impractical as so few strains produce colicins.

For several *Shigella* species, multiple single-colony isolates from a given strain will vary considerably in their colicin sensitivity (SAMARAKI-LYBEROPOULOU and PAPAVASSILIOU, 1965). The nature of these rapid changes in sensitivity is not known, but they presumably account for the unsatisfactory nature of colicinotyping for studying *Shigella* epidemics.

However, colicinotyping may be useful for epidemiological studies of *E. coli* [HAMON, 1958 (1), 1961; MCGEACHIE, 1965] as in general they are stable during a given infection or if derived from a single epidemiological focus.

Most of the type colicins will act on some *Klebsiella* strains, although 94% of
*Klebsiella* strains were not sensitive to any of these colicins [MARESZ-BABCZYSZYN,
MROZ-KURPIELA and SLOPEK, 1967 (1)]. However, nearly 40% were sensitive to one
or more of a group of colicins produced by *Shigella* strains [MARESZ-BABCZYSZYN,
MROZ-KURPIELA and SLOPEK, 1967 (2)], and in another study 60% were sensitive to
colicin A-CA31 (DE GRAAF and STOUTHAMER, 1971).

## Problems of Classifying the Bacteriocins of the Enterobacteriaceae

As might be expected with a group of bacteria which have always been difficult
to classify, it is not easy to classify the bacteriocins they produce into discrete groups
each based on a particular genus or species group.

The colicins have been well studied and we noted above that those produced by
*Salmonella*, *Shigella* or *Citrobacter* may have different activity spectra to those of *E. coli*,
although those active on *E. coli* can usually be placed within FREDERICQ's classification as types A, E, I, B, D or K. Other groups of bacteriocins such as alveicins,
cloacins and caratovoricins are also known to act on the *E. coli* and *Shigella* strains
commonly used as colicin indicators, but no cross-resistance studies have been reported as yet. They may be no more distinct from the colicins of *E. coli* than are the
colicins of *Shigella* and *Salmonella*.

It should be noted that HAMON and PERON [1966 (2, 4, 7)] consider the bacteriocins of the different *Shigella* spp. and of *Salmonella* and *Citrobacter* to constitute several
new families of bacteriocins. However, in view of the variation in spectra of the
bacteriocins of a given type within the suggested new families, it seems preferable,
for the moment at least, to retain them all in the colicin family until the heterogeneity
of the colicins has been more fully resolved. To split up the colicin family would in
part be a reversal of the amalgamation of several colicins to give the current type E
(the *Shigella* colicins were all in separate types before), and for this reason also it
seems preferable to retain the *status quo* to avoid confusion.

The marcescins of *Serratia* and the pneumocins of *Klebsiella* seem to both comprise
two groups, one specific to the particular genus and the other closely resembling the
colicins and fitting into the colicin classification. We have already referred to the
marcescins which resemble colicins of type E or K, and to the pneumocins which
resemble colicin B, and there is also the example of PAPAVASSILIOU's identification
of "colicins" of types E, I and K produced by *Klebsiella* strains.

The evolutionary origins of the two groups of bacteriocins in these genera is not
known.

## Relationships between Bacteriocins of the Enterobacteriaceae

It has long been known that some colicins are remarkably stable to heat whereas
others are relatively easily inactivated, and HAMON et al. (1966) divide colicins into
two groups on the basis of their stability to heat and various chemical agents
(see Chap. 2). They find that stability to heat and these other agents is almost confined to colicins A, D, K, N, P and V, and that the other colicins, B, C, E, H, I and
L, are labile, as are almost all other bacteriocins. The colicins of *Salmonella* and
*Shigella* are usually heat-labile and, if typable, of types B, E or I. HAMON and PERON

[1964 (3), 1965 (4), 1966 (5)] and HAMON *et al.* (1966) draw attention to this fact, and stress that colicins of types E and I are those most frequently determined by transferable C-factors. They suggest that the bacteriocins of all the other *Enterobacteriaceae* have been derived from the colicins of *E. coli*, only those which are transferable having been able to spread to other genera. However, this does not explain why such a variety of bacteriocins should have arisen in *E. coli* but not in the other Enterobacteriaceae in the first place. Colicins have also been classified on the basis of cross-reactions of neutralising antisera [HAMON and PERON, 1965 (3), 1966 (8)]. On this basis they find colicins of types E2 and E3 to be related and colicins C, E1 and L to constitute another group. Other relationships were less easily classified. The grouping together of C, L and E1 is of interest as mutants frequently show cross-resistance between these three colicins [HAUDUROY and PAPAVASSILIOU, 1962 (1)].

The colicins then show frequent interactions; they select mutants with multiple resistance, they show cross-neutralisation, and, as we have seen in other chapters, they show interactions in the immunity their C-factors confer. These interactions sometimes extend to other bacteriocins of the Enterobacteriaceae but as yet, however, it does not seem possible to draw any satisfactory conclusions regarding their significance.

## The Pseudomonadaceae

### Pyocins
(named after *Pseudomonas pyocyanea*)

JACOB (1954) studied various properties of pyocin C10 but studies on activity spectra were not carried out until later, when it was reported that almost all strains of *P. pyocyanea* produced a bacteriocin (HAMON, 1956; HAMON, VERON and PERON, 1961). Since that time there have been several studies of this type on pyocins, some motivated by the need for a typing system for epidemiological work on *P. pyocyanea* [DARRELL and WAHBA, 1964; PAPAVASSILIOU, 1961 (3); PATERSON, 1965; WAHBA, 1963, 1965 (1, 2)]. In each case a very high proportion, often more than 90%, of strains were found to be pyocinogenic.

The activity spectra of the pyocins show considerable diversity, HAMON *et al.* (1961) finding only 2 out of 16 to have identical spectra. However, some of the spectra were very similar and the differences were perhaps due to quantitative differences in pyocin production. PAPAVASSILIOU tested 22 pyocinogenic strains on 27 *P. pyocyanea* strains and this time 18 different spectra were observed, none occurring more than three times.

There are now four schemes for typing strains of *P. pyocyanea* according to the activity spectra of the pyocins they produce, and the technique seems to be of value in epidemiology. The first scheme [DARRELL and WAHBA, 1964; WAHBA, 1963, 1965 (1, 2)] used 12 indicator strains and can distinguish 11 different pyocin types. This system has proved useful [WAHBA, 1965 (1, 2)] when used in conjunction with serological typing. However, BERGAN (1968) was unable to obtain reproducible results with these indicators.

The scheme developed by PATERSON (1965) has not yet been used on any large strain collection. She was able to divide 26 pyocinogenic strains into 4 groups containing from 3 to 13 strains each, with 6 strains not fitting into any of the 4 groups. The complete table is given and there appears to be considerable diversity within groups, particularly the large group A. Of particular interest is the almost complete absence of activity of pyocins on strains in the same group, suggesting that these strains are immune to pyocins of the same type as that produced. PATERSON's scheme used indicator strains different from WAHBA's, and the relationships of the types defined in each is not yet known.

GILLIES and GOVAN (1966) developed a scheme using 8 indicator strains. Of 3227 strains examined, 88% were typable and fell into 36 pyocin types. GOVAN and GILLIES (1969) were able to report that the scheme had been used successfully in several laboratories and that a 37th type had been discovered, and that the commonest type could be subdivided into 8 by the use of additional indicator strains.

A fourth typing scheme was described by FARMER and HERMAN (1969), who used it in conjunction with other typing procedures.

*Pseudomonas pyocyanea* clearly produces a wide variety of pyocins, but since none of the typing schemes have been critically compared, it is impossible to say how many activity spectra are being distinguished in all. Since many strains may produce more than one pyocin, some of the spectra may be the sum of one or more others and the variety of pyocins be less than at first sight. Unfortunately, resistant mutants seem not to occur (PATERSON, 1965), and this rules out the use of a classification on the same basis as that of colicins. Both PATERSON (1965) and WAHBA [1965 (1, 2)] find some correlation between pyocin types produced and other properties. Thus PATERSON finds that strains lysogenic for phages of identical spectra also produce pyocins of the same type, although there is no general correlation between pyocin type and prophage carried. WAHBA [1965 (1, 2)] finds some correlation between serotype and pyocin type, and in two cases where he found indicators to subdivide a particular pyocin type, the subtypes corresponded with serotypes.

*P. pyocyanea* have also been typed according to their sensitivity patterns to a selected group of pyocins (OSMAN, 1965). Ten types were recognized on the basis of sensitivity to 4 selected pyocins. FARMER and HERMAN (1969) also included pyocin sensitivity in their composite typing scheme.

Other species of *Pseudomonas*, in particular *P. fluorescens*, have been tested for pyocin sensitivity. *P. fluorescens* is less frequently sensitive than *P. pyocyanea* (HAMON et al., 1961). PATERSON (1965) tested her 27 pyocinogenic strains against 8 strains of *P. fluorescens* and found very little activity, 3 pyocins inhibiting one particular strain of *P. fluorescens*, the rest none at all. Thus even within the genus there is a preferential action on strains of the producing species. However HAMON (1956) does report some activity on *E. coli* strains and says that rough forms of *E. coli* are very sensitive. This is to be contrasted with the absence of colicin activity on *Pseudomonas*.

We have already seen (Chap. 2) that one pyocin is, like some colicins, associated with the cell wall and that other "pyocins" are defective phages. HAMON (1956) studied several culture supernatants containing pyocins and found them to be heat-labile and non-dialysable and some to be trypsin-sensitive. Pyocins then have properties indicating that they are proteins; the mode of action of pyocin C 10 was discussed in Chap. 6.

7*

The pyocins are one of the better-defined groups of bacteriocins and in most ways resemble colicins, although the rarity or near absence of resistant mutants suggests that there may be fundamental differences between them.

## Fluocins

(named after *Pseudomonas fluorescens*)

HAMON *et al.* (1961) found 12 out of 28 strains of this species to be bacteriocinogenic, although it should be noted that PATERSON (1965) found none out of 8 strains. However, HAMON used ultraviolet induction to detect many of the fluocins. Each of the 12 fluocins observed by HAMON had a distinct activity spectrum, and in general inhibited several *P. fluorescens* strains, but only one of them had any effect on *P. pyocyanea* strains. One of them was also active on 2 strains of *Serratia*. The fluocins then seem to be well-defined bacteriocins and HAMON's observation of heat lability (70°) is at least compatible with their being proteins although only one of them was destroyed by trypsin.

## Bacteriocins of Other Pseudomonas Species

HAMON *et al.* (1961) also included in their study 10 strains of phytopathogenic *Pseudomonas* of various species, two of which were bacteriocinogenic. One (*P. aptata*) acted on several *P. fluorescens*, one *P. pyocyanea* and 2 *Serratia* strains, while the other (*P. lachrymans*) acted on 6 strains of other phytopathogenic *Pseudomonas* species. While the latter is indicative of another family of bacteriocins, the former suggests that the borderline with fluocins may not be well defined.

## Bacteriocins of Xanthomonas

Some strains of this genus produce bacteriocins active on other *Xanthomonas* and also on *Erwinia* spp. The status of these bacteriocins is not yet clear [HAMON and PERON, 1962 (1)].

## Bacteriocins of Achromobacter

MOORE and PICKETT [1960 (1)], in a general study on this genus, found three cultures to produce a substance active on other strains of the genus, each of the three having a different activity spectrum. The activity was not transmissible and the substances resembled bacteriocins. In a second paper MOORE and PICKETT [1960 (2)] described several other strains (probably also *Achromobacter*), which produce similar substances. The proportion of producing strains was high. No other studies were carried out on these bacteriocins.

MARE and COETZEE (1964) studied bacteriocins of *Alkaligenes faecalis*. We have included them here because MOORE and PICKET in their studies included strains originally thought to be *A. faecalis*, and they in fact doubt the validity of the genus *Alkaligenes*. Without passing any judgement on the taxonomic problem, it seems simpler for us to consider the two genera together. MARE and COETZEE found that two strains of *A. faecalis* were active against many members of the same species and many strains of *Proteus* spp. One of the strains (9220) was also active against a number of *Staphylococcus* strains. The activity against both *Proteus* and *Alkaligenes* moved in the same way on agar electrophoresis, suggesting that the same agent is involved. The active materials can pass through dialysis tubing and that from 9220 is heat labile (85°, 30 min), resistant to trypsin and pepsin, and precipitable by alcohol-acetone. A large number of other strains of *A. faecalis* were reported to be inhibitory for several members of the same species as well as of *Escherichia*, *Salmonella*, *Shigella*, *Serratia*, *Staphylococcus*, gram-positive aerobic bacilli and *Proteus*. As yet no detailed activity spectra have been reported for this study. It seems possible that these substances may be bacteriocins, although such widespread activity, including gram-positive bacteria, is unusual for bacteriocins of gram-negative species and the possibility remains that they are more like the classical antibiotics.

## Vibriocins

### (named after *Vibrio comma*)

BHASKARAN (1960) in a study of genetic recombination in *Vibrio cholerae* found that donor strains, P+ by analogy with F+ strains of *E. coli*, produced clearing when plated with P− strains; it is possible that these clearings are analogous to the lacunae observed under similar situations using colicinogenic cultures. TAKEYA and SHIMODORI (1969) made similar observations on *V. cholerae* and *V. El Tor*. Neither author could demonstrate any effect with killed bacteria or extracts. One P+ strain of *V. El Tor* is able to kill other strains during cell contact on millepore membranes (IYER and BHASKARAN, 1969).

WAHBA [1965 (3)] was able to demonstrate that most *V. cholerae* and *V. El Tor* strains could produce a diffusible bacteriocin on agar but that the conditions were very critical. The bacteriocins had varied activity spectra on other *V. cholerae* and *V. El Tor* strains. DATTA and PRESCOTT (1969), using a different technique, demonstrated activity against several genera of Enterobacteriaceae.

FARKAS-HIMSLEY and SEYFRIED [1962, 1963 (1, 2), 1965] have studied a substance, produced under certain conditions by a strain of *V. cholerae*, which they call vibriocin. Although the studies illuminate an interesting phenomenon, it does not seem certain that the substance is analogous to other bacteriocins and detailed discussion would be beyond the scope of this monograph.

In no case does the status of the vibriocins seem clear.

# Other Gram-negative Bacteria

## The Pesticins

### (named after *Pasteurella pestis*)

BEN-GURION and HERTMAN (1958) described the first pesticin; since that time another type has been described (BRUBAKER and SURGALLA, 1961), and the two are known as P1 and P2. Both production of these pesticins and sensitivity to them are closely correlated with the classification of the genus into species and serotypes.

Thus pesticin P1 is produced by 90% of all strains of *P. pestis* (BEN-GURION and HERTMAN, 1958; BRUBAKER and SURGALLA, 1961). Pesticin P2 is produced by all strains of both *P. pestis* and *P. pseudotuberculosis*. The activity spectrum of pesticin P 1 shows a similar close correlation with other characters, all strains of serotype 1 of *P. pseudotuberculosis* being sensitive. In addition some strains of *P. pestis* are sensitive, including some but not all of the non-producing strains and some strains which produce P1; the latter group of strains are poor growers, presumably because of their sensitivity to the pesticin they produce. Pesticin P2, on the other hand, has an activity spectrum confined to a few strains of *P. pestis* which do not produce P1. The full interrelationships are shown in Table A.1.

It is clear that the pesticins do not fit the same pattern as the colicins but for the present at least they are included as bacteriocins because their activity is almost

Table A.1. *Interrelationships between production and sensitivity of pesticins* (*From* BRUBAKER *et al., 1965*)

| Species | Strain or Type | Production | | Sensitivity | |
|---------|----------------|-----|------|-----|-----|
|         |                | PI  | PII  | PI  | PII |
| *P. pestis* | Wild | + | + | | |
| *P. pestis* | Java |  | + | | + |
| *P. pestis* | PW 12 | + | + | + | |
| *P. pestis* | A12 |  | + | + | + |
| *P. pestis* | TRU |  | + | | |
| *P. pseudotuberculosis* | Type I |  | + | + | |
| *P. pseudotuberculosis* | Types II, III, IV, V |  | + | | |
| *E. coli* | φ | − | − | + | |
| *E. coli* | B, K12 | − | − | | |

restricted to *Pasteurella*, although very few other species have been examined for sensitivity. P1 acts on *E. coli* φ but not K12 or B (Table A.1), while P2 acts on none of these three.

The activity of *P. pestis* against *E. coli* is thought to be due to the agent P1 as it follows the P1 activity in purification studies so far (BRUBAKER, personal communication).

P1 shows some further relationship to colicins in that of 9 mutants of *E. coli* φ selected for resistance to P1, 4 had lost their sensitivity to colicins B, D, I and S1. The detailed significance of this finding is not known, however, as such a mutation need not involve receptors and as yet the nature of the resistance conferred to either colicin or pesticin is not known.

Mutants of *P. pseudotuberculosis* or *E. coli* resistant to the P1 of one strain are resistant to the P1 of all other strains (BEN-GURION and HERTMAN, 1958; SMITH and BURROWS, 1962), suggesting that pesticin P1 is a single substance regardless of source.

BRUBAKER and SURGALLA (1962) found that they could increase the activity of their preparation of P1 100- to 1000-fold by removing an inhibitor by DEAE cellulose chromatography or by acid precipitation of P1. The nature of this inhibitor is not known but its presence necessitates doing P1 assays on a medium which neutralizes the effect of inhibitor. This inhibitor may resemble that described for colicin E3-CA38 and discussed in Chap. 6 as involved in immunity.

The relationship of P1 production to virulence has been studied extensively (BRUBAKER and SURGALLA, 1961; BRUBAKER, BEESLEY and SURGALLA, 1965; BRU-BAKER, SURGALLA and BEESLEY, 1965; BURROWS, 1965). There is some correlation with virulence as all mutants of *P. pestis* which do not produce P1 are also avirulent, although some avirulent mutants still produce P1. However, the detailed studies on these various mutants fall outside the scope of this book.

The activity of P2 is dependent on $Ca^{++}$ (BRUBAKER and SURGALLA, 1961); it was found that increased $Ca^{++}$ concentration up to 0.35 M led to an increased zone of inhibition from pesticin. However, $Ca^{++}$ at high concentration is deleterious for the sensitive organisms as there was no growth at $0.6\ MCa^{++}$. To their surprise they found that *E. coli* extracts would also inhibit the sensitive strain in the presence of 0.2 M $Ca^{++}$. Furthermore, the activity was specific in that *P. pestis* A1122, resistant to P2, was not affected under the same conditions. This observation of a P2-like substance produced by *E. coli* and also by *Shigella* certainly throws doubt on the nature of pesticin 2, which seems to be a very widespread inhibitor rather than a bacteriocin.

To summarize the work on pesticins described in this and other chapters then, they resemble bacteriocins in having an activity against related species, and in that P1-resistant mutants of *E. coli* show cross-resistance to some colicins. Both P1 and P2 are probably protein and are ultraviolet inducible. Like many other bacteriocins, they differ from the colicins in that both production and sensitivity to them seem to be properties of particular species or serotypes.

## Convexins

(named after *Eggerthella convexa*, also called *Bacteroides convexus*)

Of 12 strains tested, 4 produced antibiotics active against some of the 12 strains, and 3 had identical activity spectra (BEERENS and BARON, 1965). They were thermo-labile (55—65°) and are possibly bacteriocins as they are heterogeneous with respect to activity spectra, although little else is known of them.

## Columnaricins

(named after *Chondrococcus columnaris*)

ANACKER and ORDAL (1959), while looking for a phage typing scheme for this fish pathogen, found that many strains were bacteriocinogenic and they were able to recognize 7 distinct activity spectra using other strains of the same species. The patterns described resemble those typical of bacteriocins, although no other species were included in the study and so we have no information on the extent of their activity spectra. Two of them were examined and found to be heat labile (100 °C); it seems likely that they are in fact bacteriocins.

## Meningocins

(named after *Neisseria meningitidis*)

KINGSBURY (1966) found that a high proportion (28/32) of strains of *N. meningitidis* produced substances active against other strains of the same species when tested by the double-layer technique. Six of them were tested against various strains of this genus and each had a different activity spectrum. The only one looked at in detail resembled other bacteriocins in being destroyed by trypsin and being non-dialysable. It was stable to heat.

# Bacteriocins of Gram-positive Bacteria

## Staphylococcins

(named after *Staphylococcus*)

Antibiotics produced by this genus have been studied by many people, and HAMON (1964) refers to them as staphylococcins and micrococcins but as yet the subdivision does not seem to be justified. They were among the first to be described as bacteriocins, as FREDERICQ [1946 (2)] recognized that these antibiotics were analogous to the colicins. Several authors have found that staphylococcins exhibit a wide variety of activity spectra [BARROW, 1963 (1); FREDERICQ, 1946 (2); HALBERT, SWICK and SONN, 1953; LACHOWICZ, 1962, 1965], and mutants occur which are specifically resistant to particular staphylococcins [FREDERICQ, 1946 (2); HALBERT *et al.*, 1953]. Staphylococcins vary in their heat lability, sensitivity to proteolytic enzymes and precipitation by acetone or ammonium sulphate (HALBERT *et al.*, 1953).

Antibiosis among staphylococci had been studied long before FREDERICQ's work (see FLOREY *et al.*, 1949, for review), and more recently there has been emphasis on strains active against *Corynebacterium*. These strains are mostly of phage type 71 and are also active against almost all other *Staphylococcus aureus* strains and a wide range of gram-positive species [BARROW, 1963 (1); DAJANI and WANNAMAKER, 1969; HALBERT *et al.*, 1953; PARKER and SIMMONS, 1959]. The activity spectra against *Corynebacterium*, and the appearance of the inhibition zone, suggest the existence of two substances (PARKER and SIMMONS, 1959). They are probably protein in nature since they either do not pass, or pass only slowly, through dialysis membrane, and may be destroyed by proteolytic enzymes or precipitated by ammonium sulphate [BARROW, 1963 (2); DAJANI and WANNAMAKER, 1969; GARDNER, 1949].

There have been other studies on "antibiotics" of *Staphylococcus* and many of these are referred to in the references given. However, there has been no comprehensive work to ascertain the variety produced, or the relationship between those active on *Staphyloccus* and separable into many types, and those on *Corynebacterium* and apparently occurring as two types only. While some of them are apparently bacteriocins, the status of others is not yet clear.

HOFFMAN and STREITFELD (1965) have shown that the delta hemolysin of *S. aureus* has bactericidal properties, but its relationship to the substances discussed above is not clear.

## Enterococcins

(named after *Enterococcus*)

About 50% of *Streptococcus* produce bacteriocins and these vary in their activity spectra (BRANDIS and BRANDIS, 1963; BROCK *et al.*, 1963; KELSTRUP and GIBBONS, 1969; KJEMS, 1965). In all cases the producing strain was immune to the enterococcin it produced.

BROCK *et al.* (1963) surveyed a collection of 99 enterococci of various species for bacteriocin production against 19 indicators. They were able to divide their bacteriocins into five types, with production and sensitivity generally characteristic of particular species, as can be seen in Table A.2, although the correlation is far from absolute. They consider the bacteriocins of KJEMS (1965) to resemble their types 3 and 4, but it should be noted that the activity spectra of KJEMS's strains imply that more than two bacteriocins are produced by the species *S. faecalis* and *S. faecium*. The five enterococcin types were also tested on species other than group D streptococci. The type 1 bacteriocin, which we saw in Chap. 6 to be the streptococcal group D hemolysin, acts on all of the gram-positive bacteria tested with the exception of a strain of *B. polymyxa*, but not on the two gram-negative bacteria used. Type 2 acts on Lactobacilli, other Streptococci and *Leuconostoc citrovorum* (in the same family as *Streptococcus*), while types 3 and 4 act only on the later, and type 5 seems to be restricted to group D streptococci (all the species in Table A.2 are of group D).

BRANDIS and BRANDIS (1963) obtained similar results, but there seemed to be rather more variety in activity spectra, in particular amongst the bacteriocins produced by *S. liquifaciens*, and there is no obvious correlation between bacteriocin type and the producing species. They also obtained resistant mutants which showed considerable specificity, particularly with respect to the enterococcins of *S. faecium*. KUTTNER (1966) found that certain serological types of *S. pyogenes* produced bacteriocins, but their activity spectra were not studied in detail.

Enterococcins are often labile to proteolytic enzymes (BRANDIS *et al.*, 1965; BROCK *et al.*, 1963) and there is a good correlation between lability to enzymes and species producing the enterococcin. Some of this data is included in Table A.2, together with data on chloroform and heat stability. Apart from type 1, the enterococcins seem to have their activity confined to related organisms, as is expected of bacteriocins. Their lability to heat, chloroform or proteolytic enzymes suggests that they may be protein, and they should certainly be considered as bacteriocins at present.

## Lactocins

(named after *Lactobacillus*)

Of 199 strains of *Lactobacillus* tested for bacteriocin production, 12 exhibited activity and these were tested against 355 strains of the same genus (DE KLERK and COETZEE, 1961). Three of the lactocins had identical activity spectra (inhibited 38 of 554 strains) and the others were all different and inhibited from 2 to 44 strains. Their activity was also tested against some other species but no quantitative details given. A few *Streptococcus pneumoniae* and many enterococci were susceptible to some of the 12, but the Enterobacteriaceae and staphylococci were not.

Table A.2. *The enterococcins described by* Brock *et al., 1963*

| Type | Producing strains | Sensitivity of | | | | Sensitivity to | | |
|---|---|---|---|---|---|---|---|---|
| | | S. zymogenes | S. liquifaciens | S. faecium | S. faecalis | CHCl₃ | Heat | Protease |
| 1 | All S. zymogenes | − | + | + | + | S | R | R |
| 2 | Some S. liquifaciens | − | − | + | + or − | R | S | R |
| 3 | 4/40 strains of S. faecium and S. faecalis | + | + | + or − | + or − | R | R | S |
| 4 | 2/5 strains of S. faecium | + | + | + or − | + or − | R | R | S |
| 5 | 1 strain of S. zymogenes | + or − | + or − | − | + or − | R | S | R |

In their heterogeneity of activity spectra within the genus, the lactocins seem typical bacteriocins, and their activity outside the genus seems to be confined to the family Lactobacteriaceae. The activity spectra suggest that they are related to the enterococcins.

## Monocins

### (named after *Listeria monocytogenes*)

Monocins were first observed during a study on bacteriophages when SWORD and PICKETT (1961) looked at 123 strains for lysogenic phages and, in addition to success in their objective, also found 31 strains which produced nontransmissible inhibitors with definite host ranges. This observation was not followed up at the time, but HAMON and PERON [1961 (3), 1962 (2), 1963 (3)], using a different collection of strains, found 25 of 51 strains to produce bacteriocins. Only 9 strains showed sensitivity, and on the basis of activity spectra 22 of the 25 monocins were identical; the other 3 differed only slightly. Resistant mutants conferred resistance to all but one of the 25 monocins. On the basis of their activity spectra within the species, the monocins seem not to exist in the same variety as do colicins or pyocins, although clearly several different substances are involved because they differ in heat stability.

The activity of 13 monocins was tested against other genera. All 13 strains in general acted on 34/88 *Staphylococcus*, 6/20 *Bacillus* spp. and 0/18 *Streptococcus* strains; a different selection of 7 monocins had no activity against some strains of *E. coli*, *Serratia*, *Erwinia*, *Pseudomonas* and *Xanthomonas*.

The majority of monocins, then, have an almost identical activity spectrum which extends over several unrelated genera of gram-positive bacteria. They resemble other bacteriocins in being ultraviolet inducible and heat labile. They are also precipitable with neutral salts and are neutralised by antisera. TUBYLEWICZ (1963) also studied *Listeria* and found 3 of 8 strains to be bacteriocinogenic. Two distinct activity spectra were represented. Like those of HAMON and PERON, these monocins were resistant to the action of trypsin, and the two studies seem to be in general agreement. HAMON and PERON [1966 (6)] suggest on the basis of sedimentation properties and serological cross-reaction, that the "monocins" may be defective bacteriophages. As yet, however, the evidence is circumstantial.

## Megacins

### (named after *Bacillus megaterium*)

This group of bacteriocins has been extensively studied by two groups working in Hungary and in England, and three types of megacin are recognised.

Megacin A was discovered by IVANOVICS and ALFOLDI (1954). The type producer is *B. megaterium* 216, and in a systematic study of 200 strains, 40% produced a megacin (IVANOVICS and NAGY, 1958). The technique used was rather unusual in that drops of the culture to be tested were put onto a lawn of the indicator strain on agar, irradiated, and then incubated. There was thus no preincubation of the strain to be tested for bacteriocinogeny. All the megacinogenic strains showed production of megacin and a drop in turbidity after ultraviolet irradiation of growing broth cultures. However, strain 216 produced more than the others with a titre of 1/40,000 compared to a maximum of 1/100 to 1/1,000 for the others (NAGY, 1958), using strain Mut as indicator.

NAGY, ALFOLDI and IVANOVICS (1959) compared the activity of megacin A from 6 different strains on each of 15 different indicators. The results clearly demonstrated qualitative as well as quantitative differences between the megacins. Each acted to some extent on all *B. megaterium* strains tested, although HOLLAND and ROBERTS (1964) report insensitive strains. Megacins of type A have been tested against other species of bacteria (IVANOVICS and ALFOLDI, 1954; IVANOVICS, ALFOLDI and ABRAHAM, 1955; NAGY *et al.*, 1959) and all but *B. anthracis* were completely resistant although the number of strains used was not great. Megacin A-216 was not active on *B. anthracis* while other megacins were. However, OZAKI *et al.* (1966) found that megacin A-216 is lethal for *B. anthracis* in the presence of bicarbonate but they did not try other organisms under these conditions. It seems clear that megacin A includes several different types, and this was confirmed by the observation (NAGY *et al.*, 1959) that antisera to megacin A-216 would neutralise the homologous megacin but none of 16 others tested. These types of variation have not been used as the basis of any subclassification of megacins; the absence of resistant mutants (NAGY, 1965) and the almost universal sensitivity of *B. megaterium* strains to all of them makes this more difficult than for many bacteriocins.

Megacin A producing strains have, however, been divided into two groups (HOLLAND and ROBERTS, 1964). One group lyses and produces high titres of megacin in liquid medium after ultraviolet irradiation, and produces megacin C (see below) in early log phase. These strains will also immediately kill certain *B. megaterium* strains by cell-to-cell contact. The second group of megacin A producers lyse after ultraviolet irradiation but there is no increase in megacin titre; they produce little or no megacin C and their contact killing is delayed 30—40 min. The phenomenon of contact killing was described by HOLLAND and ROBERTS (1963, 1964) and is demonstrated by all megacin A producers, in either the immediate or delayed form, against certain *B. megaterium* strains. Despite its correlation with megacin A production, contact killing is equally effective against strains resistant or sensitive to megacin A, and non-megacinogenic mutants retain their contact killing characteristics.

Megacin A-216 is a phospholipase A and its chemical nature and mode of action have been discussed in the relevant chapters. The kinetics of its synthesis after induction have also been well documented (IVANOVICS and ALFOLDI, 1957).

Megacin B is produced by several *B. megaterium* strains (HOLLAND and ROBERTS, 1964). It has a more restricted activity spectrum than megacin A and producing strains do not show contact killing. Megacin B production is not induced by ultraviolet irradiation.

Megacin C is produced in early log phase by some megacin A producers, and also mutants of these strains which no longer produce megacin A (HOLLAND and ROBERTS, 1964). No megacin B producer produced any megacin C. The activity spectra of megacins of type C were tested by the double-layer technique on peptone agar (using non-megacin A producing mutants). All 6 megacins of type C tested had identical spectra, most non-megacinogenic strains being sensitive. A limited number of other strains of both gram-negative and gram-positive species were all unaffected by megacin C. Megacin C-216 was anomalous in that it was released into the medium in only small amounts.

In 1962, MARJAI and IVANOVICS described a megacin produced by *B. megaterium* 337, and called by them KP (Killer particle) and by others Megacin Cx (DURNER and MACH, 1966). We will call it KP-337. This megacin had an activity spectrum limited to only some of the *B. megaterium* strains tested. IVANOVICS, ALFOLDI and

SZELL (1957) had previously studied lysogeny in *B. megaterium*, and only some strains were sensitive to any phage at all; it is these strains which were sensitive to KP-337. Other work on this megacin has been discussed in chapters II and VI. It is clearly distinct from megacin A-216 and resembles the megacins of type C as it is produced only by young cultures and is not inducible. In its kinetics of formation it resembles C-C4 and its mode of action may also be similar (see Chap. 6). Since different indicator strains were used, the activity spectra cannot be compared directly. They do not appear to have been the same, KP-337 being active only against "phage sensitive strains" (MARJAI and IVANOVICS, 1962) while megacin C acts on most non-megacin producing strains (HOLLAND and ROBERTS, 1964). It is for this reason alone that we have not thought it advisable as yet to call it a type C megacin.

The megacins clearly have some of the properties of bacteriocins; they are macro-molecular and affect mostly *B. megaterium* strains. However, megacin A differs from colicins at least in its mode of action.

## Cerecins
### (named after *Bacillus cereus*)

In a study on a lysogenic bacteriophage, McCLOY (1951) discovered an antibiotic produced by *B. cereus* which is active on other *Bacillus* strains. HAMON and PERON [1963 (2)] refer to a study of 8 cerecins but give no details. CSUZI and KRAMER (1962) studied a substance produced by one strain of *B. cereus* and active on another. It was sensitive to trypsin, was heat labile and precipitated by ammonium sulphate and may be a bacteriocin.

## Subtilicins
### (named after *Bacillus subtilis*)

NOMURA and HOSODA [1956 (1, 2)], following up an observation that a strain of *B. subtilis* underwent autolysis under certain conditions, found an autolysin specific for some strains of *B. subtilis* and *B. megaterium*. This substance may resemble some other bacteriocins of gram-positive bacteria which seem to be lytic enzymes.

Defective bacteriophages which resemble bacteriocins in some respects have also been discovered in *B. subtilis* (SUBBAIAH *et al.*, 1965; OKAMOTO *et al.*, 1968).

## Perfringocins
### (named after *Clostridium perfringens*)

Several species of *Clostridium* produce bacteriocins. BETZ and ANDERSON (1964) found 25 strains of *C. sporogenes* produced substances active on other strains of the same species, but activity on other species was not investigated. SASARMAN and ANTOHI (1963) found that 4 of 237 strains of various types of *C. perfringens* produced bacteriocins and TUBYLEWICZ (1966) found 5 of 35 strains of *C. perfringens* type A to produce a bacteriocin. HONGO, MURATA, KONO and KATO (1968) similarly showed that 18 of 106 non-pathogenic soil Clostridia (perhaps *C. saccharoperbutyl-acetonicum*) produced a bacteriocin. In each case the bacteriocins were produced in liquid media, sometimes after ultraviolet light induction.

TUBYLEWICZ (1966) showed that his bacteriocins had no effect on 37 strains of other genera but did not include any other species of *Clostridium*, and in the other studies no strains of other species were included. Thus it is not yet known if the bacteriocins of *C. perfringens*, *C. sporogenes*, and *C. saccharoperbutylacetonicum* have similar activity spectra or if several classes of bacteriocin are involved. Those of *C. perfringens* are perhaps best called perfringocins and the others not named until they are distinguished from perfringocins.

KAUTTNER, HARMON, LYNT and LILLY (1966) found that several strains of *Clostridium* (perhaps non toxic strains of *C. botulinum* type E) produced bacteriocins active on typical strains of *C. botulinum* type E. The activity was apparently specific although only a few other strains were tested.

## Types of Bacteriocins

The examples of antibiosis which we have discussed in this appendix obviously encompass only a fraction of the studies on this subject, and there have undoubtedly been errors of selection in both directions: some cases of antibiosis currently ascribed to bacteriocins will be found to be due to unrelated agents and other examples of antibiosis, on closer study, will be found to fit the bacteriocin pattern. Amongst those we have discussed, some have been studied in some detail and clearly show the required properties o factivity mainly on closely related species and the occurrence of a variety of types. In other cases very little is known and their inclusion as bacteriocins can only be considered provisional. One interesting generalisation can be made and is often referred to by HAMON, and that is that bacteriocins of gram-positive bacteria have no activity on the gram-negatives and vice versa. However, even this statement may need qualification as some *E. coli* can act on gram-positive bacteria (COOK et al., 1953), as can some *Alkaligenes faecalis* (MARE and COETZEE, 1964), although these are perhaps minor exceptions and may not be due to bacteriocins.

Despite the heterogeneity of the substances referred to as bacteriocins it is possible to see some order in the chaos. First there are the bacteriocins such as colicins, with activity restricted to related species, and in which the variation in ability to produce a particular type, or variation in sensitivity, is not in general correlated with other characteristics. Bacteriocins of this type include the pyocins, fluocins and pneumocins; in fact, all of the known bacteriocins of gram-negative bacteria fit this description with the exception of pesticins, although of course for some of them so little is known that it is not possible to be very certain. However, some of the gram-positive bacteria produce very different bacteriocins. While some of the staphylococcins and lactocins, in as far as they have been studied, fit the colicin pattern very closely, *Staphylococci* can also produce bacteriocins with much wider activity spectra, and at the other extreme to the colicins we have type 1 enterococcin, produced by all strains of *Streptococcus zymogenes* and active on almost all other gram-positive bacteria. This particular substance is also hemolytic and presumably acts as a bacteriocin by lysing the bacterial membrane. Between the extremes we have a wide variety of intermediate types. Megacin A-216 is a phospholipase and presumably also lyses bacterial membranes. Its activity is much more restricted, being almost confined to *B. megaterium* strains. However, the early work of NAGY et al. (1959)

showed considerable variation in the potency of megacin A from different strains for other *B. megaterium* strains. It seems that there are many different types of megacin A varying in their "activity spectra" although differing from colicins in that resistance is only relative and not absolute. The other megacins, B and C, may more closely resemble colicins, and C seems to resemble colicin E2 in its mode of action.

Different again are the enterococcins other than type 1 which we referred to above. The other types have their activity restricted to closely related species but, unlike colicins, the ability to produce or conversely to be sensitive to a particular bacteriocin is often closely correlated with the classification of the genus *Streptococcus* into species.

The pesticins resemble the colicins in having their activity restricted almost entirely to other *Pasteurella* strains, but production and sensitivity are both often closely correlated with the classification into species and serotypes and only two types are known.

It remains to be seen if many groups of bacteriocins are analogous to colicins and likewise it remains to be seen how many of the bacteriocins of the gram-positive bacteria will turn out to cause membrane disintegration. For the moment it is of interest that of the two bacteriocins which are enzymes, killing by direct membrane damage, megacin A-216 and enterococcin type 1, the former has its activity restricted almost to *B. megaterium* and *B. anthracis* while the enterococcin has the widest activity spectrum known for a bacteriocin, encompassing almost all the tested gram-positive bacteria. Whether it is profitable to retain this variety of agents in the one group, the bacteriocins, is a decision for the future. That the extreme types are very different is quite clear but as yet the dividing line between them is not.

# Glossary

*Activity Spectrum.* The particular strains or species which a given bacteriocin can inhibit. Of particular use when comparing bacteriocins of different origins.

*Arbitrary Unit (A.U.).* A unit defined by some empirical test. As a unit of a bacteriocin it is usually the amount in 1 ml of a solution at a concentration just able to give complete clearing when spotted on a lawn of sensitive cells. Owing to differences in technique the unit will differ between laboratories.

*Bacteriocin.* A general term for many groups of colicin-like "antibiotics", each produced by strains of bacteria and in general active on some other strains of the same or related species. Many different groups of bacteriocins are recognised and named after the major species involved as a producer. Thus colicins are named after *Escherichia coli* and pesticins after *Pasteurella pestis.*

*Boivin antigen.* A complex particulate material which can be extracted from the surface of many gram negative bacteria. It contains lipid, carbohydrate and protein and the carbohydrate carries the O antigenic specificity (see LUDERITZ *et al.*, 1966).

*C-factor.* An epichromosome conferring on a cell the ability to synthesise a colicin.

*Clone.* As applied to bacteria, a group of cells all derived from one parent cell by repeated division without undergoing any genetic recombination.

*Colicin.* A bacteriocin of the group produced by strains of *E. coli* and related species.

*Colicinogenotype.* A subdivision of a bacterial species based on the colicins produced.

*Colicinogeny.* The ability of a bacterial cell or strain to produce a colicin.

*Colicinotype.* A subdivision of a bacterial species based on the colicins to which it is sensitive.

*Cure.* To treat a cell such that an epichromosome is lost.

*Epichromosome.* A relatively small chromosome present in only some cells of a bacterial species, in addition to the major chromosome. Often readily transferred to another cell during conjugation. In some cases it can be inserted into the main chromosome.

*Epidemic spread.* The rapid spread of an epichromosome, such as a C-factor, through a bacterial culture. Occurs when a newly transferred C-factor rapidly renders a cell able to transfer colicinogeny to other cells. Can only occur if the C-factor can replicate in much less than a cell generation time.

*Exclusion.* The ability of a cell carrying an epichromosome (or temperate phage) to exclude a related epichromosome (or phage) from entering the cells. See NOVICK (1969).

*F+ (and F-).* Designates a bacterium or bacterial strain carrying (or not carrying) an F-factor.

*F-factor*. An epichromosome conferring on a cell the ability to transfer chromosome to a recipient. In particular the epichromosome originally carried by *E. coli* K12.

*F' (F prime)*. An F-factor which includes DNA (and hence certain genes) originally derived from the main chromosome of a bacterium.

*F' Lac, F' Gal, etc.* F'-factors carrying the genes concerned with lactose fermentation (Lac) or galactose fermentation (Gal) etc.

*F-type pilus*. A sex pilus with the same antigenic and other specifities as those carried by *E. coli* K12 F+.

*HFC*. Term applied to a culture having a high frequency of transfer of colicinogeny. The bacteria of such a culture are in a special state. Usually due to their having only recently been made colicinogenic.

*Hfr (High frequency recombination)*. Designates a bacterium or bacterial strain in which an F-factor is incorporated in the main chromosome, conferring an ability for high frequency of chromosome transfer and hence of genetic recombination.

*Immunity (to bacteriocins)*. The ability of a bacterium which produces a bacteriocin to survive, often with no obvious ill effect, the action of a similar bacteriocin which would otherwise kill it.

*Immunity (to temperate phage)*. The ability of a bacterium carrying a prophage to prevent a superinfecting similar phage from replicating.

*Incompatibility*. The inability of two epichromosomes to coinhabit the same cell stably. See NOVICK (1969).

*Induction (of a bacteriocin)*. The production of a large amount of bacteriocin by a cell after some stimulus, usually ultraviolet light or a specific organic compound.

*Inhibition zone*. A zone around a bacterial colony growing on agar in which a bacteriocin (or other agent) produced by and diffusing from the bacteria of the colony is present in sufficiently high concentration to inhibit growth in an otherwise continuous lawn of the sensitive bacteria.

*I-type pilus*. A sex pilus with the same antigenic and other specificities as those determined by the C-factor I-P9.

*Lacuna*. The zone of inhibition of a lawn of a sensitive strain, produced by the bacteriocin released from a single bacterium.

*Lethal Unit (L. U.)*. The average amount of bacteriocin required to kill a sensitive cell.

*LFC*. Term applied to a culture having a low frequency of transfer of colicinogeny (used where necessary to distinguish the culture from an HFC culture).

*Multiplicity of infection (m. o. i.)*. A term derived from virology, where it means the average number of viruses adsorbed per cell. Is extended to include average number of lethal units of bacteriocin adsorbed per cell.

*O antigen*. A very characteristic antigen present on many gram-negative bacteria. The antigen consists of polysaccharide and antigenic specificity is determined by the particular variety and arrangement of the component monosaccharides. There is a very wide variety of such specificities within such genera as *Escherichia* and *Salmonella* (see LUDERITZ et al., 1966).

*Origin (of an Hfr strain)*. The point on the bacterial chromosome, adjacent to the site of F-factor attachment, which is transferred first during conjugation. In different Hfr strains the F-factor has attached at different places, giving them different origins.

*Promotion (of chromosome or epichromosome transfer)*. A general term encompassing the steps involved in organising the transmission of a chromosome or epichromosome from one cell to another.

*Receptor (for a bacteriocin)*. The part of the cell surface to which a bacteriocin specifically adsorbs prior to killing a cell.

*Refractory*. A synonym for tolerant.

*Resistant*. This term is used in a restricted sense for mutants which resist the action of a bacteriocin because they lack a receptor and cannot adsorb it (see also "tolerant").

*Restriction*. The DNA of a bacteriophage grown on one host strain may be "restricted" and degraded if the phage infects a different bacterial strain, because the DNA is recognised as being of foreign origin. See HAYES (1968) or ARBER and LINN (1969) for a full explanation.

*Sex pilus*. A pilus, or threadlike structure projecting from the surface of a bacterium, which is specifically associated with conjugation (see "F-type pilus" and "I-type pilus").

*Tolerant*. This term is used for mutants which resist (or are tolerant of) the action of a bacteriocin despite still being able to adsorb it. (See also "resistant".)

# Bibliography

Abbott, J. D., Graham, J. M.: Colicine typing of *Shigella sonnei*. Monthly Bull. Minist. Hlth. Lab. Serv. **20**, 51—58 (1961).
— Shannon, R.: A method for typing *Shigella sonnei* using colicine production as a marker. J. Clin. Pathol. **11**, 71—77 (1958).
Alfoldi, L., Jacob, F., Wollmann, E. L., Maze, R.: Sur le déterminisme génétique de la colicinogénie. Compt. rend. **246**, 3531—3533 (1958).
Amano, J., Goebel, W. F., Miller-Smidth, E.: Colicine K. III. The immunological properties of a substance having colicine K activity. J. Exp. Med. **108**, 731—752 (1958).
Amati, P.: Vegetative multiplication of colicinogenic factors after induction in *Escherichia coli*. J. Mol. Biol. **8**, 239—246 (1964).
— Ozeki, H.: Transfer of colicinogenic factors to *Serratia marcescens*. Intern. Congr. Microbiol. 8th (1962).
Anacker, R. L., Ordal, E.: Studies on the myxobacterium *Chondrococcus columnaris*. II. Bacteriocins. J. Bacteriol. **78**, 33—40 (1959).
Aoki, Y., Naito, T., Fujise, N., Ikeda, A., Miura, K., Yakushiji, Y.: Colicin typing of *Shigella sonnei*. II. The relation between colicine typing methods of Abbott and Shannon and of the authors. Japan. J. Microbiol. **11**, 73—85 (1967).
Arber, W., Linn, S.: DNA modification and restriction. Ann. Rev. Biochem. **38**, 467—506 (1969).
Atkinson, N.: Salmonellin, a new colicin-like antibiotic. Nature **213**, 184—185 (1967).
Barbour, S. D., Clark, A. J.: Biochemical and genetic studies of recombination proficiency in *Escherichia coli*. I. Enzymatic activity associated with rec B+ and rec C+ genes. Proc. Nat. Acad. Sci. US **65**, 955—961 (1970).
Baron, L. S., Formal, S. B., Spilman, W.: Vi phage-host interactions in *Salmonella typhosa*. J. Bacteriol. **69**, 177—183 (1955).
Barrow, G. I.: (1) Microbial antagonism by *Staphylococcus aureus*. J. Gen. Microbiol. **31**, 471—481 (1963).
— (2) The nature of inhibitory activity by *Staphylococcus aureus* type 71. J. Gen. Microbiol. **32**, 255—261 (1963).
— Ellis, C.: Colicine typing of *Shigella sonnei* by replicate multiple-slide inoculation of indicator organisms. Monthly Bull. Minist. Hlth. Lab. Serv. **21**, 141—147 (1962).
— — Colicin types and antibiotic sensitivity of *Shigella sonnei* in Bradford during 1962—64. Monthly Bull. Minist. Hlth. Lab. Serv. **26**, 250—253 (1967).
Barry, G. T., Everhart, D. L., Graham, M.: Colicine A. Nature **198**, 211—213 (1963).
— — Abbot, V., Graham, M. G.: Preparation, properties and relationship of substances possessing colicine A activity obtained from enterobacteriaceae. Zentr. Bakteriol. Parasitenk. Abt. I Orig. **196**, 248—263 (1965).
Basinger, S. F., Jackson, R. W.: Bacteriocin (Hemolysin) of *Streptococcus zymogenes*. J. Bacteriol. **96**, 1895—1902 (1968).
Bayer, M. E., Anderson, T. F.: The surface structure of *Escherichia coli*. Proc. Nat. Acad. Sci. US **54**, 1592—1599 (1965).
Bazaral, M., Helinski, D. R.: (1) Circular DNA forms of colicinogenic factors E1, E2 and E3 from *Escherichia coli*. J. Mol. Biol. **36**, 185—194 (1968).
— — (2) Characterization of multiple circular DNA forms of colicinogenic factor E1 from *Proteus mirabilis*. Biochemistry **7**, 3513—3519 (1968).
Beale, G. H.: The genetics of *Paramecium aurelia*. Cambridge: University Press 1954.
— Jurand, A.: Structure of the mate-killer (mu) particles in *Paramecium aurelia*. stock 540. J. Gen. Microbiol. **23**, 243—252 (1960).

BEALE, G. H., JURAND, A.: Three different types of mate-killer (mu) in *Paramecium aurelia* (syngen 1). J. Cell. Sci. **1**, 31—34 (1966).

BEERENS, H., BARON, G.: Mise en évidence de bactériocines elaborées par les bactéries anaérobies à gram négatif appartenant au genre *Eggerthella*. Ann. inst. Pasteur **108**, 255—256 (1965).

BELAYA, O. S., EMEL'AYNOVA, O. I., VOLINA, L. E., STAVNICHAYA, A. P., TEPLITSKAYA, E. S., VAISBERG, L. A., SHMATKOVA, A., TROVITSKAYA, L. S., EGEL'SKAYA, G. V., OSADCHAYA, N. V., YANUSH, P. A., TASHCHICKINA, Z. G., OL'KHOVSKAYA, K. A.: Intraspecies and intratype microbial differentiation of Enterobacteriaceae with the aid of standard colicins. Zhur. Mikrobiol. Epidemiol. Immunobiol. **46** (6), 72—73 (1969).

BEN-GURION, R.: On the nature of the "lethal zygote" produced by crossing non-colicinogenic with colicinogenic bacteria. J. Gen. Microbiol. **30**, 173—181 (1963).

— On the induction of a recombination-deficient mutant of *Escherichia coli* K12. Genet. Res. **9**, 377—381 (1967).

— HERTMAN, I.: Bacteriocin-like material produced by *Pasteurella pestis*. J. Gen. Microbiol. **19**, 289—297 (1958).

BEPPU, T., ARIMA, K.: Protection of *Escherichia coli* from the lethal effect of colicins by high osmotic pressure. J. Bacteriol. **93**, 80—85 (1967).

— — Properties of the colicin E2 induced degradation of DNA of *E. coli*. J. Biochem. Tokyo. **70**, 263—271 (1971).

BERGAN, T.: Typing of *Pseudomonas aeruginosa* by pyocine typing. Acta Pathol. Microbiol. Scand. **72**, 401—411 (1968).

BERNSTEIN, A., ROLFE, B., ONODERA, K., TILL, J. E.: Genetic fine structure and pleiotropic characteristics of the TolAB locus of *E. coli*. Bacteriol. Proc. **1971**, 50.

BERTANI, G.: Infections bactériophagiques secondaires des bactéries lysogènes. Ann. inst. Pasteur **84**, 273—280 (1953).

BETZ, J. V., ANDERSON, K. E.: Isolation and characterisation of bacteriophages active on *Clostridia sporogenes*. J. Bacteriol. **87**, 408—415 (1964).

BHASKARAN, K.: Recombination of characters between mutant stocks of *Vibro cholerae*, strain 162. J. Gen. Microbiol. **23**, 47—54 (1960).

BHATTACHARYYA, P., WENDT, L., WHITNEY, E., SILVER, S.: Colicin-tolerant mutants of *Escherichia coli*: resistance of membranes to colicin E. Science **168**, 998—1000 (1970).

BILLEN, D.: Replication of the bacterial chromosome: location of new initiation sites after irradiation. J. Bacteriol. **97**, 1169—1175 (1969).

BLACKFORD, V. L., PARR, L. W., ROBBINS, M. L.: Antibiotic activity of selected enteric organisms. Antibiotics Chemotherapy **1**, 392—398 (1951).

BOIVIN, A., MESROBEANU, L.: Extraction d'un complexe toxique et antigénique à partir du bacile d'Aertryche. Compt. rend. soc. biol. **114**, 307—310 (1933).

— — Recherches sur les antigènes somatiques et sur les endotoxines des bactéries. I. Considérations générales et exposé des techniques utilisées. Rev. immunol. **1**, 553—569 (1935).

BOON, T.: Inactivation of ribosomes *in vitro* by colicin E3. Proc. Natl. Acad. Sci. US **68**, 2421—2425 (1971).

BORDET, P.: Propriétés distinctes d'antibiotiques produits, par deux souches de colibacille dont l'une derive de l'autre. Rev. immunol. **11**, 323 (1947).

— Inhibition de l'action d'antibiotiques par les sérums antibactériens. Compt. rend. soc. biol. **142**, 257 (1948).

— BEUMER, J.: Inhibition de l'action d'antibiotiques par des extraits des bactéries sensibles. Compt. rend. soc. biol. **142**, 259—261 (1948).

— — Antibiotiques et bactériophages. Bull. acad. roy. méd. Belg. **14**, 116—134 (1949).

— — Antibiotiques colibacillaires et récepteurs appropriés. Rev. belge pathol. et méd. exptl. **21**, 245—250 (1951).

BOWMAN, C. M., DAHLBERG, J. E., IKEMURA, T., KONISKY, J., NOMURA, M.: Specific inactivation of 16S ribosomal RNA induced by colicin E3 *in vivo*. Proc. Natl. Acad. Sci. US **68**, 964—968 (1971).

— SIDIKARO, J., NOMURA, M.: Specific inactivation of ribosomes by colicin E3 *in vitro* and mechanism of immunity in colicinogenic cells. Nature New Biology **234**, 133—137 (1971).

BRADLEY, D. E.: Ultrastructure of bacteriophages and bacteriocins. Bacteriol. Rev. 31, 230—314 (1967).
— DEWAR, C. A.: The structure of phage-like objects associated with non-induced bacteriocinogenic bacteria. J. Gen. Microbiol. 45, 399—408 (1966).
BRANCHE, W. C., YOUNG, V. M., ROBINET, C. G., MASSEY, E. D.: Effect of colicine production on Escherichia coli in the normal human intestine. Proc. Soc. Exp. Biol. Med. 114, 198—201 (1963).
BRANDIS, H., BRANDIS, U.: Auftreten und Verhalten spontaner Mutanten von Enterokokkenstämmen mit Resistenz gegen Enterocin. Pathol. Microbiol. 26, 688—695 (1963).
— — VAN DE LOO, W.: Bacteriocinähnliche Stoffe von Enterokokken. Zentr. Bakteriol. Parasitenk. Abt. I. Orig. 196, 331—346 (1965).
BRAUDE, A. I., SIEMIENSKI, J.: Plasma bactericidal power of mice injected with non-toxic colicin V. Federation Proc. 23, 565 (1964).
— SIEMIENSKI, J. S.: The influence of bacteriocins on resistance to infection by gram negative bacteria. I. The effect of colicin on the bactericidal power of blood. J. Clin. Invest. 44, 849—859 (1965).
BRENNER, S., JACOB, F., MESELSON, M.: An unstable intermediate carrying information from genes to ribosomes for protein synthesis. Nature 190, 576—581 (1961).
BRINTON, C. C.: The structure, function, synthesis and genetic control of bacterial pili and a molecular model for DNA and RNA transport in gram negative bacteria. Trans. N.Y. Acad. Sci. 27, 1003—1054 (1965).
BROCK, T. D., DAVIE, J. M.: Probable identity of a group D hemolysin with a bacteriocine. J. Bacteriol. 86, 708—712 (1963).
— PEACHER, B., PIERSON, D.: Survey of the bacteriocins of enterococci. J. Bacteriol. 86, 702—707 (1963).
BROOKS, K., CLARK, A. J.: Behaviour of λ bacteriophage in a recombination deficient strain of E. coli. J. Virol. 1, 283—293 (1967).
BRUBAKER, R. R., BEESLEY, E. D., SURGALLA, M. J.: Pasteurella pestis. Role of the Pesticin I and iron in experimental plague. Science 149, 422—424 (1965).
— SURGALLA, M. J.: Pesticins. I. Pesticin-bacterium interrelationships, and environmental factors influencing activity. J. Bacteriol. 82, 940—949 (9161).
— — Pesticins. II. Production of pesticins I and II. J. Bacteriol. 84, 539—545 (1962).
— — BEESLEY, E. D.: Pesticinogeny and bacterial virulence. Zentr. Bakteriol. Parasitenk. Abt. I. Orig. 196, 302—315 (1965).
BURROWS, T. W.: A possible role for pesticin in virulence of Pasteurella pestis. Zentr. Bakteriol. Parasitenk. Abt. I. Orig. 196. 315—317 (1965).
BUTTIN, G.: Some aspects of regulation in the synthesis of the enzymes governing galactose metabolism. Cold Spring Harbor Symposia Quant. Biol. 26, 213—216 (1961).
CAIRNS, J.: A minimum estimate of the length of the DNA of Escherichia coli obtained by autoradiography. J. Mol. Biol. 4, 407—409 (1962).
CARO, L. G., SCHNOS, M.: The attachment of male specific phage f1 to sensitive strains of Escherichia coli. Proc. Nat. Acad. Sci. US 56, 126—132 (1966).
CAVARD, D., BARBU, E.: Quelques données sur la fixation des colicines. Compt. rend. 262, 1809—1812 (1966).
— — Evaluation du nombre des récepteurs bactériens pour les colicines A, E1, E2 et K. Ann. inst. Pasteur 119, 420—431 (1970).
— POLONOVSKI, J., BARBU, E.: Modifications du métabolisme des phospholipides de E. coli consécutives à l'action de la colicine K. Compt. rend. 265D, 1851—1854 (1967).
CEFALU, M., BAVASTRELLI, L.: Sensibilità alle colicine e proprietà colicinogene nel gruppo Shigella. Boll. inst. sieroterap. milan. 38, 86—98 (1959).
CHAKRABARTY, A. N.: Observations on the colicinogenity of Shigella flexneri. Ind. J. Med. Research 52, 1226—1230 (1964).
CHANGEUX, J. P., THIERY, J.: On the mode of action of colicins: model of regulation at the membrane level. J. Theoret. Biol. 17, 315—318 (1967).
CHATTERJEE, A. N.: Use of bacteriophage-resistant mutants to study the nature of the bacteriophage receptor site of Staphylococcus aureus. J. Bacteriol. 98, 519—527 (1969).

CHATTERJEE, A. N., MIRELMAN, D., SINGER, H. J., PARK, J. T.: Properties of a novel pleiotropic bacteriophage resistant mutant of *Staphylococcus aureus* H. J. Bacteriol. **160**, 846—853 (1969).

CLARK, A. J., MARGUILES, A. D.: Isolation and characterisation of recombination deficient mutants of *E. coli* K12. Proc. Nat. Acad. Sci. US **53**, 451—459 (1965).

CLEWELL, D. B., HELINSKI, D. R.: Existence of the C-factor Col Ib-P9 as a supercoiled DNA protein relaxation complex. Biochem. Biophys. Res. Commun. **41**, 150—156 (1970).

CLOWES, R. C.: Colicine factors as fertility factors in bacteria: *Escherichia coli* K12. Nature **190**, 988—989 (1961).

— Colicin factors and episomes. Genet. Res. Camb. **4**, 162—165 (1963).

— Transfert génétique des facteurs colicinogènes. Ann. inst. Pasteur **107** (suppl. to No. 5), 74—92 (1964).

— Transmission and elimination of colicin factors and some aspects of immunity to colicin E1 in *Escherichia coli*. Zentr. Bakteriol. Parasitk. Abt. I. Orig. **196**, 152—160 (1965).

— HAUSMANN, C., NISIOKA, T., MITANI, M.: Genetic analysis of Col B Factors and the identification of composite circular molecules of R-factors. J. Gen. Microbiol. **55**, iv. (1969).

— MOODY, E. E. M., PRITCHARD, R. H.: The elimination of extrachromosomal elements in thymineless strains of *Escherichia coli* K12. Genet. Res. **6**, 147—152 (1965).

COCITO, C., VANDERMEULEN-COCITO, J.: Properties of a colicine, selectively inhibiting the growth of *E. coli* B and favouring multiplication of mutants resistant to phage T3, in static and continuous cultures. Giorn. microbiol. **6**, 146—179 (1958).

COETZEE, J. N.: Transmission of colicinogeny to *Providence* strains. Nature **203**, 897—898 (1964).

— Bacteriocinogeny in strains of *Providence* and *Proteus morganii*. Nature **213**, 614—616 (1967).

COOK, G. T., DAINES, C. F.: Colicine types of *Shigella sonnei* in West Surrey. Monthly Bull. Minist. Hlth. Lab. Serv. **23**, 81—85 (1964).

COOK, M. K., BLACKFORD, V. L., ROBBINS, M. L., PARR, L. W.: An investigation of the antibacterial spectrum of colicins. Antibiotics & Chemotherapy **3**, 195—202 (1953).

COOK, T. M., DEITZ, W. H., GOSS, W. A.: Mechanism of action of nalidixic acid on *Escherichia coli*. IV. Effects on the stability of cellular constituents. J. Bacteriol. **91**, 774—779 (1966).

CRADDOCK-WATSON, J. E.: The production of bacteriocines by *Proteus* species. Zentr. Backteriol. Parasitenk. Abt. I. Orig. **196**, 385—388 (1965).

CSUZI, S., KRAMER, M.: Production of a lytic factor by ultraviolet light irradiated cultures of *Bacillus cereus* I. The conditions of lysis induction. Acta Microbiol. Acad. Sci. Hung. **9**, 297—304 (1962).

CURTISS, R., CHARAMELLA, L. J., STALLIONS, D. R., MAYS, J. A.: Parental functions during conjugation in *Escherichia coli* K12. Bacteriol. Rev. **32**, 320—348 (1968).

DAJANI, A. S., WANNAMAKER, L. W.: Demonstration of a bactericidal substance against β-hemolytic streptococci in supernatant fluids of staphylococcal cultures. J. Bacteriol. **97**, 985—991 (1969).

DANDEU, J. P., BARBU, E.: Purification de la colicine K. Compt. rend. **265D**, 774—776 (1967).

— — Étude comparée de quelques colicines. Compt. rend. **266D**, 634—636 (1968).

DARRELL, J. H., WAHBA, A. H.: Pyocin-typing of hospital strains of *Pseudomonas pyocyanea*. J. Clin. Path. **17**, 236—242 (1964).

DATTA, A., PRESCOTT, L. M.: Effect of vibriocins on members of the enterobacteriaceae J. Bacteriol. **98**, 849—850 (1969).

DAVIE, J. M., BROCK, T. D.: (1) Action of streptolysin S, the group D hemolysin, and phospholipase C on whole cells and spheroplasts. J. Bacteriol. **91**, 595—600 (1966).

— — (2) Effect of Teichoic acid on resistance to the membrane-lytic agent of *Streptococcus zymogenes*, J. Bacteriol. **92**, 1623—1631 (1966).

DEITZ, W. H., COOK, T. M., GOSS, W. A.: Mechanism of action of nalidixic acid on *Escherichia coli*. III. Conditions required for lethality. J. Bacteriol. **91**, 768—773 (1966).

DE PETRIS, S.: Ultrastructure of the cell wall of *Escherichia coli* and chemical nature of its constituent layers. J. Ultrastruct. Research **19**, 45—83 (1967).

Depoux, R., Chabbert, Y.: Étude de l'activé antibiotique d'une colicine. Ann. inst. Pasteur **84**, 798—801 (1953).

De Witt, W., Helinski, D. R.: Characterization of colicinogenic factor E1 from a non-induced and a mitomycin C-induced *Proteus* strain. J. Mol. Biol. **13**, 692—703 (1965).

Dowman, J. E., Meynell, G. G.: Pleiotropic effects of de-repressed bacterial sex-factors on colicinogeny and cell wall structure. Mol. Gen. Genet. **109**, 57—68 (1970).

Driskell-Zamenhof, P. J., Adelberg, E. A.: Studies on the chemical nature and size of sex-factors of *Escherichia coli*. J. Mol. Biol. **6**, 483—497 (1963).

Durlakowa, I., Lachowicz, Z.: Transmission of factors conditioning bacteriocinogeny in *Klebsella* bacilli. Arch. Immunol. Therap. Exp. **15**, 540—546 (1967).

— Maresz-Babczyszyn, J., Przondo-Hessek, A., Lusar, Z., Mroz-Kurpiela, E.: (1) The phenomenon of bacteriocinogeny in bacilli of the genus K*lebsiella*. I. Characteristics of *Klebsiella* bacteriocins. Arch. Immunol. Therap. Exp. **12**, 308—318 (1964).

— — — — — (2) The phenomenon of bacteriocinogeny in bacilli of the genus *Klebsiella*. II. Bacteriocins produced by strains of *Klebsiella* bacilli isolated from patients with ozena. Arch. Immunol. Therap. Exp. **12**, 319—331 (1964).

Durner, K., Mach, F.: Physiologische Untersuchungen eines Bacteriocin aus *Bacillus megaterium* 337. Zentr. Bakteriol. Parasitenk. Abt. II **120**, 565—575 (1966).

Echols, H.: Properties of *Escherichia coli* superinfected with F-lactose and F-galactose episomes. J. Bacteriol. **85**, 262—268 (1963).

Edwards, S., Meynell, G. G.: General method for isolating de-repressed bacterial sex factors. Nature **219**, 869—870 (1968).

Elgat, M., Ben-Gurion, R.: Mode of action of pesticin. J. Bacteriol. **98**, 359—367 (1969).

Endo, H., Kamiya, T., Ishizawa, M.: $\lambda$ phage induction by colicine E2. Biochem. Biophys. Res. Commun. **11**, 477—482 (1963).

— Ayabe, K., Amano, K., Takeya, K.: Inducible phage of *Escherichia coli* 15. Virology **25**, 469—471 (1965).

Farkas-Himsley, H., Seyfried, P. L.: Lethal biosynthesis of a bacteriocin, vibriocin. Nature **193**, 1193—1194 (1962).

— — (1) Lethal biosynthesis of a bacteriocin, Vibriocin, by *V. comma*. I. Conditions affecting its production and detection. Can. J. Microbiol. **9**, 329—338 (1963).

— — (2) Lethal biosynthesis of a bacteriocin, Vibriocin, by *V. comma*. II Vibriocin production and sensitivity in relation to redox potentials and streptomycin resistance. Can. J. Microbiol. **9**, 339—343 (1963).

— — Vibriocin and nucleic acids. Zentr. Bacteriol. Parasitenk. Abt. I. Orig. **196**, 298—302 (1965).

Farmer, J. J., Herman, L. G.: Epidemiological fingerprinting of *Pseudomonas aeruginosa* by the production of and sensitivity to pyocin and bacteriophage. Appl. Microbiol. **18**, 760—765 (1969).

Favre, R., Amati, P., Bezdek, M.: Properties of P22 and a related *Salmonella typhimurium* phage. Virology **35**, 238—247 (1968).

Fields, K. L.: Comparison of the action of colicins E1 and K on *Escherichia coli* with the effects of abortive infection by virulent bacteriophages. J. Bacteriol. **97**, 78—82 (1969).

— Luria, S. E.: (1) Effects of colicins E1 and K on transport systems. J. Bacteriol. **97**, 57—63 (1969).

— — (2) Effects of colicins E1 and K on cellular metabolism. J. Bacteriol. **97**, 64—77 (1969).

Filichkin, S. E.: Colicin sensitivity of Flexner and Sonne dysentery bacteria and the practical value of the method of colicin sensitivity determination. Zhur. Mikrobiol. Epidemiol. Immunobiol. **45**, 69—75 (1968).

Florey, H. M., Chain, E., Heatley, N. G., Jennings, M. A., Sanders, A. G., Abraham, E. P., Florey, M. E.: The antibiotics. Oxford: University Press (1949).

Foulds, J. D., Shemin, D.: Concomitant synthesis of bacteriocin and bacteriocin inactivator from *Serratia marcescens*. J. Bacteriol. **99**, 661—666 (1969).

Frampton, E. W., Brinkley, B. R.: Evidence of lysogeny in derivatives of *Escherichia coli*. J. Bacteriol. **90**, 446—452 (1965).

FREDERICQ, P.: (1) Sur la classification des *B. coli* d'après leurs caractères antibiotiques. Compt. rend. soc. biol. **140**, 1055—1057 (1946).

— (2) Sur la sensibilité et l'activité antibiotique des staphylocoques. Compt. rend. soc. biol. **140**, 1167—1170 (1946).

— (3) Sur la pluralité des récepteurs d'antibiose de *E. coli*. Compt. rend. soc. biol. **140**, 1189—1190 (1946).

— (4) Action bactéricide de la colicine K. Compt. rend. soc. biol. **140**, 1295—1296 (1946).

— (5) Sur la specificité des actions antibiotiques. Schweiz. Z. allgem. Pathol. Bakeriol. **9**, 385—390 (1946).

— Actions antibiotiques réciproques chez les Enterobacteriaceae. Rev. belge path. et méd. exp. **19**. suppl. 4, 1—107 (1948).

— (1) Sur la résistance croisée entre colicines E et bactériophage II. Compt. rend. soc. biol. **143**, 1011—1013 (1949).

— (2) Sur la résistance croisée entre colicin K et bactériophage III. Compt. rend. soc. biol. **143**, 1014—1017 (1949).

— (1) Recherches sur les colicines du groupe E. Compt. rend. soc. biol. **144**, 297—299 (1950).

— (2) Un phénomène d'antibiose simulant les taches de bactériophagie. Compt. rend. soc. biol. **144**, 730—732 (1950).

— (3) Recherches sur les caractères et la distribution des souches productrices de colicine B. Compt. rend. soc. biol. **144**, 1237—1298 (1950).

— (4) Acquisition de propriétés antibiotiques nouvelles par une souche d'*E. coli* sous l'action de certains bactériophages. Compt. rend. soc. biol. **144**, 1709—1711 (1950).

— (1) Induction spécifiqué de propriétés antibiotiques nouvelles sous l'effet de certains bactériophages. Compt. rend. soc. biol. **145**, 141—143 (1951).

— (2) Recherches sur l'origine des mutants de *E. coli* V produisant la colicine M. Compt. rend. soc. biol. **145**, 930—933 (1951).

— (3) Spectres d'activité différents des souches productrices de colicine E ou K et des bactériophages des groupes II ou III. Compt. rend. soc. biol. **145**, 1433—1436 (1951).

— (4) Acquisition de propriétés antibiotiques nouvelles par la souche *E. coli* V sous l'action des bactériophages T₁, T₅ et T₇. Antonie van Leeuwenhoek. J. Microbiol. Serol. **17**, 102—106 (1951).

— (5) Origine spontanée des mutants de *E. coli* V produisant la colicine M. Antonie van Leeuwenhoek. J. Microbiol. Serol. **17**, 227—231 (1951).

— (1) Recherches des propriétés lysogènes et antibiotiques chez les *Salmonella*. Compt. rend. soc. biol. **146**, 298—300 (1952).

— (2) Action antibiotique d'un bactériophage. Bull. soc. chim. biol. **35**, 415—417 (1952).

— (1) Recherches sur les caractères et la distribution des souches productrices de diverses colicines dans les selles normales et pathologiques. Bull. acad. roy. méd. Belg. **18**, 126—139 (1953).

— (2) Action des colicins E et K sur la multiplication de divers bactériophages. Compt. rend. soc. biol. **147**, 533—537 (1953).

— (1) Transduction génétique des propriétés colicinogènes chez *Escherichia coli* et *Shigella sonnei*. Compt. rend. soc. biol. **148**, 399—402 (1954).

— (2) Induction de la production de colicine par irradiation ultraviolette de souches colicinogène d'*Escherichia coli*. Compt. rend. soc. biol. **148**, 1276—1280 (1954).

— Induction de la production de colicine et de bactériophages par irradiation ultraviolette de souches colicinogènes et lysogènes d'*Escherichia coli*. Compt. rend. soc. biol. **149**, 2028—2031 (1955).

— (1) Recherches sur la fréquence des souches transductrices des propriétés colicinogènes. Compt. rend. soc. biol. **150**, 1036—1039 (1956).

— (2) Résistance et immunité aux colicines. Compt. rend. soc. biol. **150**, 1514—1517 (1956).

— Colicins. Ann. Rev. Microbiol. **11**, 7—22 (1957).

— (1) Transduction of colicinogenicity. In Department of Genetics report. Carnegie Inst. Yearbook **57**, 396—397 (1958).

— (2) Colicins and colicinogenic factors. Symposium Soc. Exp. Biol. **12**, 104—122 (1958).

— Transduction par bactériophage des propriétés colicinogènes chez *Salmonella typhimurium*. Compt. rend. soc. biol. **153**, 357—360 (1959).

FREDERICQ, P.: (1) Colicines et autres bactériocines. Ergeb. Mikrobiol. Immunitätsforsch. u. exp. Therap. **37**, 114—161 (1963).
— (2) Linkage of colicinogenic factors with F-agents and with the chromosome in *E. coli.* Microbial. Genetics Bull. **19**, 9—10 (1963).
— (3) On the nature of colicinogenic factors: a review. J. Theoret. Biol. **4**, 159—165 (1963).
— Colicines et colicinogénie. Ann. inst. Pasteur **107**, (Suppl. to No. 5), 7—17 (1964).
— (1) A note on the classification of colicins. Zentr. Bakteriol. Parasitenk. Abt. I. Orig. **196**, 140—151 (1965).
— (2) Genetics of colicinogenic factors. Zentr. Bakteriol. Parasitenk. Abt. I. Orig. **196**, 142—151 (1965).
— The recombination of colicinogenic factors with other episomes and plasmids. Ciba Found. Symp. Bacterial Plasmids Episomes, 1969.
— BETZ-BARREAU, M.: (1) Présence régulière et constante d'*Escherichia freundii* dans l'urine des malades atteints de fièvre typhoïde. Compt. rend. soc. biol. **142**, 1078—1080 (1948).
— — (2) Caractères distinctifs des *Escherichia freundii* isolées de l'urine des malades typhiques. Compt. rend. soc. biol. **142**, 1180—1181 (1948).
— — Caractères d'activités et de sensibilité antibiotiques de la flore coliforme dominants de selles normales et pathologiques. Compt. rend. soc. biol. **144**, 1424—1427 (1950).
— — (1) Transfert génétique de la propriété colicinogène chez *E. coli.* Compt. rend. soc. biol. **147**, 1110—1112 (1953).
— — (2) Transfert génétique de la propriété de produire un antibiotique. Compt. rend. soc. biol. **147**, 1653—1656 (1953).
— — (3) Transfert génétique de la propriété colicinogène en rapport avec la polarité F des parents. Compt. rend. soc. biol. **147**, 2043—2045 (1953).
— — NICOLLE, P.: Typage de souches d'*Escherichia coli* de gastro-enterite infantile par recherche de leur propriétés colicinogènes. Compt. rend. soc. biol. **150**, 2039—2042 (1956).
— GRATIA, A.: Résistance croisée à certaines colicines et à certains bactériophages. Compt. rend. soc. biol. **143**, 560—563 (1949).
— — Rapports entre colicines et bactériophages du groupe T.1—T.7. Antonie van Leeuwenhoek. J. Microbiol. Serol. **16**, 119—121 (1950).
— JOIRIS, E.: Distribution des souches productrices de colicine V dans les selles normales et pathologiques. Compt. rend. soc. biol. **144**, 435—437 (1950).
— — BETZ-BARREAU, M., GRATIA, A.: Recherche des germes producteurs de colicines dans les selles des malades atteints de fièvre paratyphoïde B. Compt. rend. soc. biol. **143**, 556—559 (1949).
— SMARDA, J.: Complexité du facteur colicinogène B. Ann. inst. Pasteur **118**, 767—774 (1970).
FRIEDBERG, E. C., GOLDTHWAITE, D. A.: Endonuclease II of *E. coli.* I. Isolation and purification. Proc. Natl. Acad. Sci. US **62**, 934—940 (1969).
— HADI, S., GOLDTHWAITE, D. A.: Endonuclease II of *E. coli.* II. Enzyme properties and studies on the degradation of alkylated and native DNA. J. Biol. Chem. **244**, 5879—8889 (1969).
FUERST, G. R., SIMINOVITCH, L.: Characterisation of an unusual defective lysogenic strain of *Escherichia coli* K12 (λ). Virology **27**, 449—451 (1965).
FUJIMURA, R. K.: Effect of colicine E2 on the biosynthesis of bacteriophage R17. J. Mol. Biol. **17**, 75—85 (1966).
FURNESS, G., ROWLEY, D.: The presence of the transmissible agent F in non-recombining strains of *E. coli.* J. Gen. Microbiol. **17**, 550—561 (1957).
GARDNER, J. F.: An antibiotic produced by *Staphylococcus aureus.* Brit. J. Exp. Pathol. **30**, 130—138 (1949).
— Some antibiotics formed by *Bacterium coli.* Brit. J. Exp. Pathol. **31**, 102—111 (1950).
GAREN, A., PUCK, T. T.: The first two steps in the invasion of host cells by bacterial viruses II. J. Exp. Med. **94**, 177—189 (1951).
GERMAINE, G. R., ROGERS, P.: Role of gal repressor depletion in λ dg transduction escape synthesis. J. Mol. Biol. **47**, 121—135 (1970).
GHUYSEN, J.-M., STROMINGER, J. L., TIPPER, D. J.: Bacterial Cell Walls. In Comprehensive Biochemistry. **26A**, 53—104. Ed. Florkin and Stotz. Elsevier 1968.

GILLIES, R. R.: Colicine production as an epidemiological marker of *Shigella sonnei*. J. Hyg. **62**, 1—9 (1964).

— Bacteriocin production: an epidemiologic marker of *Sh. sonnei*: inter-regional type incidence. Zentr. Bakteriol. Parasitenk. Abt. I. Orig. **196**, 370—377 (1965).

— GOVAN, J. R. W.: Typing of *Pseudomonas pyocyanea* by pyocine production. J. Pathol. Bacteriol. **91**, 339—345 (1966).

GODARD, C., BEUMER-JOCHMANS, M. P., BEUMER, J.: Apparition de sensibilité aux phages T et à des colicines chez *Shigella flexneri* F 6 S survivant à l'infection par un phage Lisbonne. I. Modification des propriétés biologiques de surface. Ann. inst. Pasteur **120**, 475—489 (1971).

GOEBEL, W. F.: The chromatographic fractionation of colicine K. Proc. Natl. Acad. Sci. US **48**, 214—219 (1962).

— BARRY, G. T.: Colicine K. II. The preparation and properties of a substance having colicine K activity. J. Exp. Med. **107**, 185—209 (1958).

— — SHEDLOVSKY, T.: Colicine K. I. The production of colicine K in media maintained at constant pH. J. Exp. Med. **103**, 577—588 (1956).

— JESAITIS, M. A.: The somatic antigen of a phage-resistant variant of phase II *Shigella sonnei*. J. Exp. Med. **96**, 425—438 (1952).

— — Chemical and antiviral properties of the somatic antigens of phase II *Sh. sonnei* and of a phage-resistant variant 11/3, 4, 7. Ann. inst. Pasteur **84**, 66—72 (1953).

GOSS, W. A., DEITZ, W. H., COOK, T. M.: Mechanism of action of nalidixic acid on *Escherichia coli*. II. Inhibition of deoxyribonucleic acid synthesis. J. Bacteriol. **89**, 1068—1074 (1965).

GOUPILLE, F., HAMON, Y., VIEU, J.-F.: Étude de quelques souches de coliformes responsables de la mastite bovine. Ann. inst. Pasteur **98**, 577—585 (1960).

GOVAN, J. R. W., GILLIES, R. R.: Further studies in the pyocine typing of *Pseudomonas pyocyanea*. J. Med. Microbiol. **2**, 17—25 (1969).

DE GRAAF, F. K., GOEDVOLK-DE GROOT, L. E., STOUTHAMER, A. H.: Purification of a bacteriocin produced by *Enterobacter cloacae* DF13. Biochim. et Biophys. Acta **221**, 566—575 (1970).

— SPANJAERDT SPECKMAN, E. A., STOUTHAMER, A. H.: Mode of action of a bacteriocin produced by *Enterobacter cloacae*. Antonie van Leeuwenhoek. J. Microbiol. Serol. **35**, 287—306 (1969).

— STOUTHAMER, A. H.: Mode of action of a bacteriocin produced by *Enterobacter cloacae*. J. Gen. Microbiol. **53**, xiii (1969).

— — Isolation and properties of bacteriocin-tolerant mutants of *Klebsiella edwardsii* var. *edwardsii*. Antonie van Leeuwenhoek. J. Microbiol. Serol. **36**, 217—226 (1970).

— — Interaction of various bacteriocins with *Klebsiella edwardsii* var. *edwardsii*. Antonie van Leeuwenhoek. J. Microbiol. Serol. **37**, 1—14 (1971).

— TIEZE, G. A., BONGA, S. W., STOUTHAMER, A. H.: Purification and genetic determination of bacteriocin production in *Enterobacter cloacae*. J. Bacteriol. **95**, 631—640 (1968).

GRATIA, A.: Sur un remarquable exemple d'antagonisome entre deux souches de colibacille. Compt. rend. soc. biol. **93**, 1040—1041 (1925).

— Antagonisme microbien et bactériophagie. Ann. inst. Pasteur **48**, 413—437 (1932).

— FREDERICQ, P.: Diversité des souches antibiotiques de *B. coli* et étendue variable de leur champ d'action. Compt. rend. soc. biol. **140**, 1032—1033 (1946).

GRATIA, J. P.: Résistance à la colicine B chez *E. coli*. Relations de spécificité entre colicines B, I, V et phage T1, Étude génétique. Ann. inst. Pasteur **107** (Suppl. to No. 5), 132—151 (1964).

GROSS, J. D.: Cellular damage associated with multiple mating in *E. coli*. Genet. Res. **4**, 463—469 (1963).

GUTTERMAN, S. K.: Inhibition of colicin B by enterochelin. Biochem. Biophys. Res. Commun. **44**, 1149—1155 (1971).

— LURIA, S. E.: *Escherichia coli*: Strains that excrete an inhibitor of colicin B. Science **164**, 1414 (1969).

DE HAAN, P. G., STOUTHAMER, A. H.: F-prime transfer and multiplication of sexduced cells. Genet. Res. **4**, 30—41 (1963).

HALBERT, S. P.: (1) The antagonism of coliform bacteria against *Shigellae*. J. Immunol. **58**, 153—167 (1948).
— (2) The relation of antagonistic coliform organisms to *Shigella* infections. J. Immunol. **60**, 23—36 (1948).
— (3) The relation of antagonistic coliform organisms to *Shigella* infections. II. Observations in acute infections. J. Immunol. **60**, 359—381 (1948).
— MAGNUSON, H. J.: Studies with antibiotic — producing strains of *Escherichia coli*. J. Immunol. **58**, 397—415 (1948).
— SWICK, L., SONN, C.: Characteristics of antibiotic-producing strains of the occular bacterial flora. J. Immunol. **70**, 406—410 (1953).
HAMON, Y.: Étude d'une colicine elaborée par une culture de *S. paratyphi* B. Ann. inst. Pasteur **88**, 193—204 (1955).
— Contribution à l'étude des pyocines. Ann. inst. Pasteur. **91**, 82—90 (1956).
— Propriétés générales des cultures d'enterobacteriacées rendues colicinogènes par transfert. Ann. inst. Pasteur **92**, 363—368 (1957).
— (1) Étude des images de sensibilité des *Escherichia coli* pathogènes pour le nourisson à diverse colicines types. Intérêt épidémiologique de cette étude. Ann. inst. Pasteur **95**, 117—121 (1958).
— (2) Étude du pouvoir colicinogène parmi les *Escherichia coli* pathogènes pour l'adulte et l'enfant. Compt. rend. **247**, 1260—1261 (1958).
— Recherche et identification de la propriété colicinogène parmi les différents lysotypes d'*Escherichia coli* pathogènes pour le nourisson. Ann. inst. Pasteur **96**, 614—629 (1959).
— Colicines et lysotypie. Zentr. Bakteriol. Parasitenk. Abt. I. Orig. **181**, 456—468 (1961).
— Les bactériocines. Ann. inst. Pasteur **107** (Suppl. to No. 5), 18—53 (1964).
— Les bactériocines et substances analogues. Pathol. et biol. Semaine hop. **13**, 806—824 (1965).
— CHABBERT, Y. A.: Relations entre facteurs colicinogènes et facteurs de resistance chez une souche de *S. panama*. Compt. rend. **271D**, 259—262 (1970).
— MARESZ, J., HSI, T., PERON, Y.: Sur le pouvoir de conjugaison conféré aux bactéries par certains types de facteur pneumocinogènes et aérocinogènes. Compt. rend. **265D**, 1089—1091 (1967).
— — KAYSER, A., PERON, Y.: Relations de parenté de quelques aérocines de la fraction I avec certains types de colicine. Defectivité de ces facteurs aérocinogène Compt. rend. **271D**, 1713—1716 (1970).
— — PERON, Y.: Étude de l'action de quelques agents physiques, chimiques et biologiques sur les bactériophages et les bactériocines. Colloquium über Fragen der Lysotypie. Wernigerade (Harz) 1966.
— NICOLLE, P.: Sur un facteur colicinogène propre à certains lysotypes de *Salmonella typhi*. Compt. rend. **255**, 2690—2692 (1962).
— PERON, Y.: Étude du mode de fixation des colicines et des pyocines sur les bactéries sensibles. Compt. rend. **251**, 1840—1842 (1960).
— — (1) Étude de la propriété bactériocinogène dans le genre *Serratia*. Ann. inst. Pasteur **100**, 818—821 (1961).
— — (2) Les propriétés antagonistes réciproques parmi les *Erwinia*. Discussion de la position taxonomique de ce genre. Compt. rend. **253**, 913—915 (1961).
— — (3) Étude du pouvoir bactériocinogène dans le genre *Listeria*. Compt. rend. **253**, 1883—1885 (1961).
— — (1) Les bactériocines, éléments taxonomiques éventuels pour certains bactéries. Compt. rend. **254**, 2868—2870 (1962).
— — (2) Étude du pouvoir bactériocinogènes dans le genre *Listeria*. I. Propriétés générales de ces bactériocines. Ann. inst. Pasteur **103**, 876—889 (1962).
— — (3) Sur la cinétique de la libération de leur antibiotique par les divers types de microbes bactériocinogènes. Compt. rend. **255**, 2210—2212 (1962).
— — (4) Sur la libération de leurs bactériophage et de leurs antibiotiques de divers types d'*E. coli* à la fois lysogènes et colicinogènes. Arch. roumaines pathol. exptl. microbiol. **21**, 342—350 (1962).

HAMON, Y., PERON, Y.: (1) Individualisation de quelques nouvelles familles d'enterobactériocines. Compt. rend. **257**, 309—311 (1963).

— — (2) Quelques remarques sur les bactériocines produites par les microbes Grampositifs. Compt. rend. **257**, 1191—1193 (1963).

— — (3) Étude du pouvoir bactériocinogène dans le genre *Listeria*. II. Individualité et classification des bactériocines en cause. Ann. inst. Pasteur **104**, 55—65 (1963).

— — (4) Étude du pouvoir bactériocinogènes dans le genre *Cloaca*. Ann. inst. Pasteur **104**, 127—131 (1963).

— — (1) A propos de quelques nouveaux types de colicines thermostables. Compt. rend. **258**, 3121—3124 (1964).

— — (2) Contribution à l'étude des bactériocines des *Salmonella*. Compt. rend. **258**, 4162—4165 (1964).

— — (3) Quelques propriétés générales des enterobactériocins. Compt. rend. **259**, 1270—1273 (1964).

— — (4) Description de sept nouveaux types de colicines. État actuel de la classification de ces antibiotiques. Ann. inst. Pasteur **106**, 44—54 (1964).

— — (1) Essai de classification de quelques marcescines. Compt. rend. **260**, 5401—5404 (1965).

— — (2) Étude des propriétés inductrices de certaines colicines. Compt. rend. **260**, 5948—5951 (1965).

— — (3) La sérologie des bactériocines. Importance de cette methode. Compt. rend. **260**, 6730—6733 (1965).

— — (4) Étude de l'action de quelques agents physiques et chimiques sur les enterobactériocines. Compt. rend. **261**, 591—594 (1965).

— — (1) Nouvelle classification des colicines appartenant au groupe E. Zentr. Bakteriol. Parasitenk. Abt. I. Orig. **200**, 375—379 (1966).

— — (2) Contribution à l'étude de la propriété bactériocinogène dans la tribu des Salmonellae. I. Les bactériocines des *Salmonella*. Ann. inst. Pasteur **110**, 389—402 (1966).

— — (3) Relations de quelques marcescines actives sur *E. coli* avec certains types de colicines. Ann. inst. Pasteur **110**, 556—561 (1966).

— — (4) Étude des bactériocines produites par les bactéries appartenant au groupe *Citrobacter-Bethesda*. Ann. inst. Pasteur **111**, 497—501 (1966).

— — (5) Étude des relations des colicines avec les diverses enterobactériocines. Compt. rend. **262D**, 581—584 (1966).

— — (6) Sur la nature des bactériocines produites par *Listeria monocytogenes*. Compt. rend. **263D**, 198—200 (1966).

— — (7) Les bactériocines du groupe *Akalescens-Dispar* et des *Shigella*. Compt. rend. **263D**, 573—575 (1966).

— — (8) Études des relations antigéniques des colicines. Compt. rend. **263D**, 1173—1175 (1966).

— — (1) La production de queues de phages par une bactérie lysogène deféctive. Ann. inst. Pasteur **112**, 241—244 (1967).

— — (2) Étude d'une souche bactériocinogène et lysogène d'*Enterobacter aerogenes*. Zentr. Bakteriol. Parasitenk. Abt. I. Orig. **203**, 184—189 (1967).

— VERON, M., PERON, Y.: Contribution à l'étude des propriétés lysogènes et bactériocinogènes dans le genre *Pseudomonas*. Ann. inst. Pasteur. **101**, 738—753 (1961).

HART, J.: Colicine production as an epidemiological tool. Zentr. Bakteriol. Parasitenk. Abt. I. Orig. **196**, 364—369 (1965).

HASKELL, E. H., DAVERN, C. I.: Prefork synthesis: A model for DNA replication. Proc. Natl. Acad. Sci. US **64**, 1065—1071 (1969).

HAUDUROY, P., PAPAVASSILIOU, J.: (1) Identification of a new type of colicine (colicine L). Nature **195**, 730—732 (1962).

— — (2) De l'hétérogénéité des souches d'*Escherichia coli* productrices de colicine. Ann. inst. Pasteur. **102**, 644—649 (1962).

HAYES, W.: The Genetics of bacteria and their viruses, 2nd edition. Blackwell 1968.

HEATLEY, N. G., FLOREY, H. W.: An antibiotic from *Bacterium coli*. Brit. J. Exp. Pathol. **27**, 378—390 (1946).

HEDGES, A. J.: An examination of single-hit and multi-hit hypotheses in relation to the possible kinetics of colicin adsorption. J. Theoret. Biol. **11**, 383—410 (1966).

HELINSKI, D. R., HERSCHMAN, H. R.: Effect of Rec⁻ mutations on the activity of colicinogenic factors. J. Bacteriol. **94**, 700—706 (1967).

HERRIOTT, R. M., BARLOW, J. L.: The protein coats or "ghosts" of coli phage T2. II. The biological functions. J. Gen. Physiol. **41**, 307—331 (1957).

HERSCHMAN, H. R., HELINSKI, D. R.: (1) Comparative study of the events associated with colicin induction. J. Bacteriol. **94**, 691—699 (1967).

— — (2) Purification and Characterisation of colicin $E_2$ and $E_3$. J. Biol. Chem. **242**, 5360—5368 (1967).

HERTMAN, I., BEN-GURION, R.: A study of pesticin biosynthesis. J. Gen. Microbiol. **21**, 135—143 (1958).

— LURIA, S. E.: Transduction studies on the role of a rec + gene in the ultraviolet induction of prophage λ. J. Mol. Biol. **23**, 117—133 (1967).

HEWITT, R., BILLEN, D.: Reorientation of chromosome replication after exposure to ultraviolet light of *Escherichia coli*. J. Mol. Biol. **13**, 40—53 (1965).

VAN HEYNINGEN, W. E.: Bacterial toxins. Oxford: Blackwell 1950.

HIGASHI, Y., SAITO, H., YANAGASE, Y., YONEMASO, K., AMANO. T.: Studies on the role of plakin. XI. Demonstration of phospholipase A in plakin. Bikens. J. **9**, 249—262 (1966).

HIGERD, T. B., BAECHLER, C. A., BERK, R. S.: In vitro and in vivo characterization of pyocin. J. Bacteriol. **93**, 1976—1986 (1967).

HILL, C., HOLLAND, I. B.: Genetic basis of colicin E susceptibility in *Escherichia coli*. I. Isolation and properties of refractory mutants and the preliminary mapping of their mutations. J. Bacteriol. **94**, 677—686 (1967).

HINDSDILL, R. D., GOEBEL, W. F.: The chemical nature of bacteriocins. Ann. inst. Pasteur **107** (Suppl. to No. 5), 54—66 (1964).

— — Colicine K. VII. The transfer of type K colicinogeny to *Shigella sonnei*. J. Exp. Med. **123**, 881—895 (1966).

HIROTA, Y.: The effect of acridine dyes on mating type factors in *Escherichia coli* K12. Proc. Natl. Acad. Sci. US **46**, 57—64 (1960).

HOFFMANN, E. M., STREITFELD, M. M.: The antibiotic activity associated with preparations of delta hemolysin of *Staphylococcus aureus*. Can. J. Microbiol. **11**, 203—211 (1965).

HOLLAND, E. M., HOLLAND, I. B.: Induction of DNA breakdown and inhibition of cell division by colicin E2. Nature of some early steps in the process and properties of the E2-specific nuclease system. J. Gen. Microbiol. **64**, 223—239 (1970).

HOLLAND, I. B.: The purification and properties of megacin, a bacteriocin from *Bacillus megaterium*. Biochem. J. **78**, 641—648 (1961).

— Further observations on the properties of megacin, a bacteriocin formed by *Bacillus megaterium*. J. Gen. Microbiol. **29**, 603—614 (1962).

— Effect of a bacteriocin preparation (megacin C) on DNA synthesis in *Bacillus megaterium*. Biochem. Biophys. Res. Commun. **13**, 246—250 (1963).

— A bacteriocin specifically affecting DNA synthesis in *Bacillus megaterium*. J. Mol. Biol. **12**, 429—438 (1965).

— The properties of uv sensitive mutants of *Escherichia coli* K12 which are also refractory to colicin E2. Mol. Gen. Genet. **100**, 242—251 (1967).

— Properties of *Escherichia coli* K12 Mutants which show conditional refractivity to colicin E2. J. Mol. Biol. **31**, 267—275 (1968).

— ROBERTS, C. F.: A search for the transmission of megacinogenic factor in *Bacillus megaterium*. J. Gen. Microbiol. **31**, 11—12 (1963).

— — Some properties of a new bacteriocin formed by *Bacillus megaterium*. J. Gen. Microbiol. **35**, 271—285 (1964).

— THRELFALL, E. J.: Identification of closely linked loci controlling ultraviolet sensitivity and refractivity to colicin E2 in *Escherichia coli*. J. Bacteriol. **97**, 91—96 (1969).

— — HOLLAND, E. M., DARBY, V., SAMSON, A. C. R.: Mutants of *Escherichia coli* with altered surface properties, which are refractory to colicin E2, sensitive to ultraviolet light and which can also show recombination deficiency, abortive growth of bacteriophage λ and filament formation. J. Gen. Microbiol. **62**, 371—382 (1970).

Homma, J. Y., Goto, S., Shionoya, H.: Relationship between pyocine and temperate phage of *Pseudomonas aeruginosa*. II. Isolation of pyocines from strain P1-111 and their characteristics. Japan. J. Exp. Med. **37**, 373—393 (1967).

— Hamamura, N., Naoi, M., Egami, F.: Recherches chimiques sur l'endotoxine de *Pseudomonas aeruginosa* II. Bull. soc. chim. biol. **60**, 647—664 (1958).

— Shionoya, H.: Relationship between pyocine and temperate phage of *Pseudomonas aeruginosa* III. Serological relationship between pyocines and temperate phages. Japan. J. Exp. Med. **37**, 395—421 (1967).

— — Meguro, M., Tanabe, Y.: A short communication on pyocine 28 produced by *Pseudomonas aeruginosa*. Japan. J. Exp. Med. **37**, 511—513 (1967).

— Suzuki, N.: "Cell-wall protein A" of *Pseudomonas aeruginosa* and its relationship to original endotoxin protein. J. Bacteriol. **87**, 630—646 (1964).

— — The protein moiety of the endotoxin of *Pseudomonas aeruginosa*. Ann. N.Y. Acad. Sci. **133**, 508—526 (1966).

— — Ito, F.: The surface structure of *Pseudomonas aeruginosa*. J. Immunol. **90**, 819—828 (1963).

Hongo, M., Murata, A., Kono, K., Kato, F.: Lysogeny and bacteriocinogeny in strains of *Clostridium* species. Agric. Biol. Chem. **32**, 27—33 (1968).

Huet, M., Papavassiliou, J., Bonnefous, S.: Recherches sur les *Shigella* colicinogènes. Arch. inst. Pasteur **38**, 109—119 (1961).

Hull, R. R.: The mode of action of colicin. Ph. D. Thesis, University of Adelaide 1971.

— Reeves, P.: Sensitivity of intracellular bacteriophage λ to colicin CA42-E2. J. Virology **8**, 355—362 (1971).

Hutton, J. J., Goebel, W. F.: Colicine V. Proc. Natl. Acad. Sci. US **47**, 1498—1500 (1961).

— — The isolation of colicine V and a study of its immunological properties. J. Gen. Physiol. **45** (Suppl.), 125—141 (1962).

Iijima, T.: Studies on the colicinogenic factor in *Escherichia coli* K12. Induction of colicin production by mitomycin C. Bikens. J. **5**, 1—8 (1962).

Ikeda, H., Tomizawa, J.: Prophage P1, an extrachromosomal replication unit. Cold Spring Harbor Symposia Quant. Biol. **33**, 791—879 (1968).

Ishii, S., Nishi, Y., Egami, F.: The fine structure of a pyocin. J. Mol. Biol. **13**, 428—431 (1965).

Ivanovics, G.: Bacteriocins and bacteriocin-like substances. Bacteriol. Rev. **26**, 108—118 (1962).

— Alfoldi, L.: A new antibacterial principle: Megacin. Nature. **174**, 465 (1954).

— — Observation on lysogenesis in *Bacillus megaterium* and on megacin, the antibacterial principle of this *Bacillus* species. Acta Microbiol. Acad. Sci. Hung. **2**, 275—292 (1955).

— — Bacteriocinogenesis in *Bacillus megaterium* J. Gen. Microbiol. **16**, 522—530 (1957).

— — Abraham, E.: Das antibakterielle spektrum des megacins. Zentr. Bacteriol. Parasitenk. Abt. 1. Orig. **163**, 274—280 (1955).

— — Nagy, E.: Mode of action of megacin. J. Gen. Microbiol. **21**, 51—60 (1959).

— — Szell, A.: Serological types of *Bacillus megaterium* and their sensitivity to phages. Acta Microbiol. Acad. Sci. Hung. **4**, 333—351 (1957).

— Nagy, E.: Hereditary aberrancy in growth of some *Bacillus megaterium* strains. J. Gen. Microbiol. **19**, 407—418 (1958).

— — Alfoldi, L.: Megacinogeny: inducible synthesis of a new immunospecific substance. Acta Microbiol. Acad. Sci. Hung. **6**, 161—169 (1959).

Iyer, S. S., Bhaskaran, K.: A lethal factor in a strain of *Vibrio El Tor*. Genet. Res. **14**, 9—12 (1969).

Jacob, F.: Biosynthèse induit et mode d'action d'une pyocine, antibiotique de *Pseudomonas pyocyanea*. Ann. inst. Pasteur **86**, 149—160 (1954).

— Lwoff, A., Siminovitch, A., Wollman, E.: Définition de quelques termes relatifs à la lysogénie. Ann. inst. Pasteur **84**, 222—224 (1953).

— Siminovitch, L., Wollman, E.: Sur la biosynthèse d'une colicine et sur son mode d'action. Ann. inst. Pasteur **83**, 295—315 (1952).

JACOB, F., WOLLMAN, E. L.: Les épisomes, éléments génétiques ajoutés. Compt. rend. **247**, 154—156 (1958).
— — Sexuality and the genetics of bacteria. New York: Academic Press, Inc. 1961.
JAYAWARDENE, A., FARKAS-HIMSLEY, H.: Particulate nature of vibriocin: a bacteriocin from *Vibrio comma*. Nature **219**, 79—80 (1968).
JESAITIS, M. A.: Properties of colicine K produced by *Proteus mirabilis*. Bacteriol. Proc. **1967**, 50.
— The nature of colicin K from *Proteus mirabilis*. J. Exp. Med. **131**, 1016—1038 (1970).
— GOEBEL, W. F.: Mechanism of phage action. Nature **172**, 622—623 (1953).
KABAT, E. A., MAYER, M. M.: Experimental immunochemistry, 2nd edition. Springfield, Illinois: C. C. Thomas 1961.
KAGEYAMA, M.: Studies of a pyocin. I. Physical and chemical properties. J. Biochem. (Tokyo) **55**, 49—53 (1964).
— EGAMI, F.: On the purification and some properties of a pyocin, a bacteriocin produced by *Pseudomonas aerigunosa*. Life Sci. **9**, 471—476 (1962).
— IKEDA, K., EGAMI, F.: Studies of a pyocin III. Biological properties of the pyocin. J. Biochem. (Tokyo) **55**, 59—64 (1964).
KAHN, P. L.: Isolation of high-frequency recombining strains from *Escherichia coli* containing the V colicinogenic factor. J. Bacteriol. **96**, 205—214 (1968).
— HELINSKI, D. R.: Relationship between colicinogenic factors E1 and V and an F factor in *Escherichia coli*. J. Bacteriol. **88**, 1573—1579 (1964).
— — Interaction between colicinogenic factor V and the integrated F factor in an Hfr strain of *Escherichia coli*. J. Bacteriol. **90**, 1276—1282 (1965).
KALCKAR, H. M., LAURSEN, P., RAPIN, A. M. C.: Inactivation of phage C21 by various preparations from Lipopolysaccharide of *E. coli* K12. Proc. Natl. Acad. Sci. US **56**, 1852—1858 (1966).
KAMEDA, M., HARADA, K., MATSUYAMA, T.: Colicin typing and resistance pattern of R factors of *Shigella sonnei*. Japan, J. Bacteriol. **23**, 345—347 (1968).
KASATYIA, S., HAMON, Y.: Étude du pouvoir colicinogène parmi les *E. coli* 0119: B14. Rev. Hyg. et Med. Soc. **13**, 35—48 (1965).
KATES, M.: Bacterial Lipids. Advances in Lipid. Res. **2**, 17—90 (1964).
KATO, Y., HANAOKA, M.: The elimination of a colicinogenic factor by a drug-resistance transferring factor in *Escherichia coli* K235. Biken's J. **5**, 77—86 (1962).
KAUTTNER, D. A., HARMON, S. M., LYNT, R. K., LILLY, T.: Antagonistic effect on *Clostridium botulinum* type E by organisms resembling it. Appl. Microbiol. **14**, 616—622 (1966).
KAYSER, A., HAMON, Y., LAMBLIN, L.: Sur la fertilité d'une souche bactériocinogène d'*Enterobacter aerogenes* et des souches derivées. Ann. inst. Pasteur **120**, 491—500 (1971).
KAZIRO, Y., TANAKA, M.: (1) Studies on the mode of action of pyocin I. Inhibition of macromolecular synthesis in sensitive cells. J. Biochem. (Tokyo) **57**, 689—695 (1965).
— — (2) Studies on the mode of action of pyocin. II. Inactivation of ribosomes. J. Biochem. (Tokyo) **58**, 357—363 (1965).
KEENE, J. H.: Preparation and chemical properties of colicine I. J. Microbiol. **12**, 425—427 (1966).
KELSTRUP, J., GIBBONS, R. J.: Bacteriocins from human and rodent streptococci. Arch. Oral Biol. **14**, 248—251 (1969).
KINGSBURY, D. T.: Bacteriocin production by strains of *Neisseria meningitidis*. J. Bacteriol. **91**, 1696—1699 (1966).
KJEMS, E.: Studies on streptococcal bacteriophages I. Technique of isolating phage producing strains. Acta. Pathol. Microbiol. Scand. **36**, 433—440 (1955).
— DE KLERK, H. C., COETZEE, J. N.: Antibiosis among lactobacilli. Nature **192**, 340—341 (1961).
KOHIYAMA, M., NOMURA, M.: DNA synthesis and induction of colicine E2 as studied with a temperature sensitive mutant of colicinogenic *E. coli* strain. Zentr. Bakteriol. Parasitenk. Abt. 1. Orig. **196**, 211—215 (1965).
KONISKY, J.: Ph. D. Thesis, Wisconsin University 1967.
— NOMURA, M.: Interaction of colicins with bacterial cells. II specific alteration of *Escherichia coli* ribosomes induced by colicins E3 *in vivo*. J. Mol. Biol. **26**, 181—195 (1967).

KUDLAI, D. G., LIKHODED, V. G., GOLUBEVA, I. V.: Correlation between type of colicine produced and antigenic structure of pathogenic *Escherichia coli*. Zhwr. Mikrobiol. Epidemiol. Immunobiol. **41** (9), 65 (1964).

KUTTER, E. M., WIBERG, J. S.: Degradation of cytosine-containing bacterial and bacteriophage DNA after infection of *Escherichia coli* B with bacteriophage T4 D and wild type and with mutants defective in genes 46, 47 and 56. J. Mol. Biol. **388**, 395—5411 (1968).

KUTTNER, A. G.: Production of bacteriocines by group A streptococci with special reference to the nephritogenic types. J. Exp. Med. **124**, 279—291 (1966).

LACHOWICZ, T.: A study of the isoantagonism of staphylococci. I. The isolation of an antibiotic factor from solid media. Med. Doswiadczalna i Mikrobiol. **14** (1) 17—25 (1962).

— Investigations on staphylococcins. Zentr. Bacteriol. Parasitenk. Abt. 1. Orig. **196**, 340—351.

— WALCZAK, Z.: Preliminary report on purification of staphylococcin. Postepy Microbiol. **5** (2), 213—217 (1966).

LASZLO, V. G., KEREKES, L.: Phage and colicin typing of *Shigella sonnei*. Acta Microbiol. Acad. Sci. Hung. **16**, 309—317 (1969).

LAWN, A. M.: Morphological features of the pili associated with *Escherichia coli* K12 carrying R-Factors or the F-factor. J. Gen. Microbiol. **45**, 377—383 (1966).

— MEYNELL, G. G., MEYNELL, E., DATTA, N.: Sex pili and the classification of sex factors in the Enterobacteriaceae. Nature **216**, 343—346 (1967).

LEDERBERG, J.: Cell genetics and hereditary symbiosis. Physiol. Rev. **32**, 403—430 (1952).

— TATUM, E. L.: Gene recombination in *E. coli*. Nature **158**, 558 (1946).

LEHMAN, I. R., HERRIOTT, R. M.: The protein coats or ghosts of coliphage T2. III. metabolic studies of *Escherichia coli* B infected with T2 bacteriophage ghosts. J. Gen. Physiol. **41**, 1067—1082 (1958).

LE MINOR, S.: Conversions antigénique chez les salmonellas. XII Structure antigénique O des *Salmonella* du groupe W. Modifications appartées par la conversion bactériophagique. Ann. inst. Pasteur **116**, 474—483 (1969).

LEONOVA, C. V.: *Staphylococcus* bacteriocins and their properties. Zhur. Mikrobiol. Epidemiol. Immunobiol. **45** (3), 108—111 (1968).

LEVIN, A. P., BURTON, K.: Inhibition of enzyme formation following infection of *Escherichia coli* with phage T2r+. J. Gen. Microbiol. **25**, 307—314 (1961).

LEVISOHN, R., KONISKY, J., NOMURA, M.: Interaction of colicin with bacterial cells. IV. Immunity breakdown studied with colicins Ia and Ib. J. Bacteriol. **96**, 811—821 (1968).

LEWIS, M. J., STOCKER, B. A. D.: Properties of some group E colicine factors. Zentr. Bakteriol. Parasitenk. Abt. 1. Orig. **196**, 173—183 (1965).

LIKHODED, V. G.: (1) Ultraviolet induction of colicin synthesis by *E. coli*. Zhu. Mikrobiol. Epidemiol. Immunobiol. **40** (7), 116—120 (1963).

— (2) Some properties of *E. coli* colicins. Antibiotiki **8** (9), 771—777 (1963).

LINDBERG, A. A., HOLME, T.: Influence of O side chains on the attachment of the Felix O-1 bacteriophage to *Salmonella* bacteria. J. Bacteriol. **90**, 513—519 (1969).

LINTON, K. B.: The colicine typing of coliform bacilli in the study of cross-infection in urology. J. Clin. Pathol. **13**, 168—172 (1960).

LORKIEWICZ, Z., DERYLO, M., FRELIK, M.: Studies on colicinogenic factors transfer and elimination. Proc. Symp. Mutational Process. Czech. Acad. Sci. Prague (1965).

— MACIAZEK, K., NACHIEWICZ, Z.: The influence of acriflavine on transfer of the colicinogenic factor. Acta Microbiol. Polan. **13**, 273—281 (1964).

LOSICK, R., ROBBINS, P. W.: The receptor site for a bacterial virus. Sci. American. **121**, No. 5, 120—125 (1969).

LUDERITZ, O., JANN, K., WHEAT, R.: Somatic and capsular antigens of gram-negative bacteria. In: Comprehensive Biochemistry 26A, 165—228. (FLORKIN, STOTZ, Eds.) Elsevier 1968.

— STAUB, M., WESTPHAL, O.: Immunochemistry of O and R antigens of *Salmonella* and related Enterobacteriaceae. Bacteriol. Rev. **30**, 192—255 (1966).

LURIA, S. E.: On the mechanism of action of colicins. Ann. inst. Pasteur **107** (Suppl. to No. 5), 67—73 (1964).

Luzzati, D., Chevalier, M. R.: Induction, par carence en thymine, de la production de colicine par des bactéries colicinogènes thymine-exigeants. Ann. inst. Pasteur **107** (Suppl. to No. 5), 152—162 (1964).

Macfarren, A. C., Clowes, R. C.: A comparative study of two F-like colicin factors, Col V2 and Col V3 in *Escherichia coli* K12. J. Bacteriol. **94**, 365—377 (1967).

Maeda, A., Nomura, M.: Interaction of colicins with bacterial cells. I. Studies with radio-active colicins. J. Bacteriol. **91**, 685—694 (1966).

Mandel, M., Mohn, F.: Colicins in *Serratia marcescens.* Microbial Genetics Bull. **18**, 15 (1962).

Mare, I. J., Coetzee, J. N.: Antibiotics of *Alcaligens faecalis.* Nature **203**, 430—431 (1964).

Maresz, J., Durlakowa, I., Hamon, Y., Peron, Y.: Essai de classification des pneumocines. Relations de ces antibiotiques avec certains types des colicines. Compt. rend. **263D**, 85—88 (1966).

— Hamon, Y., Peron, Y.: Propriétés générales des aérocines; les bactéries aérocinogènes défectives. Compt. rend. **267D**, 2044—2047 (1968).

Maresz-Babczyszyn, J., Durlakowa, I., Lachowicz, Z., Hamon, Y.: Characteristics of bacteriocins produced by *Klebsiella* bacilli. Arch. Immunol. Ther. Exp. **15**, 530—539 (1967).

— — Mroz-Kurpiela, E., Lachowicz, Z., Slopek, S.: Sensitivity of *Klebsiella* bacilli to bacteriocins produced by bacilli of homologous group. Arch. Immunol. Ther. Exp. **15**, 521—524 (1967).

— Mroz-Kurpeila, E., Slopek, S.: (1) Sensitivity of *Klebsiella* bacilli to Fredericq's set of colicins. Arch. Immunol. Ther. Exp. **15**, 512—516 (1967).

— — — (2) Sensitivity of *Klebsiella* bacilli to the Abbott-Shannon set of colicins. Arch. Immunol. Ther. Exp. **15**, 517—520 (1967).

Marjai, E. H., Ivanovics, G.: A second bacteriocin-like principle of *Bacillus megaterium.* Acta Microbiol. Acad. Sci. Hung. **9**, 285—295 (1962).

— — The effect of different anticancer agents on inducible systems of *Bacillus megaterium.* Acta. Microbiol. Acad. Sci. Hung. **11**, 193—198 (1964).

Marotel-Schirman, J., Barbu, E.: Quelques donnés sur les récepteurs des colicins. Compt. rend. **269D**, 866—869 (1969).

Martin, H. H.: Bacterial protoplasts—a review. J. Theoret. Biol. **5**, 1—34 (1963).

Matsushita, H., Fox, M. S., Goebel, W. F.: Colicine K. IV. The effect of metabolites upon colicine synthesis. J. Exp. Med. **112**, 1055—1068 (1960).

Mayne, E.: Essai d'élimination par l'acridine organe ou la mitomycine d'un facteur colicinogène lié à un facteur de fertilité chez *Escherichia coli.* Ann. inst. Pasteur **109**, 154—157 (1965).

Mayr-Harting, A.: Nature of colicin receptors. Nature **192**, 191 (1961).

— The adsorption of colicine. J. Pathol. Bacteriol. **87**, 255—266 (1964).

— Shimeld, C.: Some observations on colicine receptors. Zentr. Bakteriol. Parasitenk. Abt. 1. Orig. **196**, 263—270 (1965).

McCloy, E. W.: Studies on a lysogenic *Bacillus* strain I.A bacteriophage specific for *Bacillus anthracis.* J. Hyg. **49**, 114—125 (1951).

McGeachi, J.: Bacteriocin typing in urinary infection. Zentr. Bakteriol. Parasitenk. Abt. 1 Orig. **196**, 377—384 (1965).

Menningman, H. D.: On the nature of inducible anti-bacterial agent of *E. coli* 15. Zentr. Bakteriol. Parasitenk. Abt. 1. Orig. **196**, 207—210 (1965).

Mesrobeanu, I., Mesrobeanu, L., Croitoresco, I., Mitrica, N.: L'activité bactériocinique des neurotoxines des germes Gram-négatifs. Compt. rend. **258**, 1919—1921 (1964).

Meynell, E. W.: A phage $\phi$ x, which attacks motile bacteria. J. Gen. Microbiol. **25**, 253—290 (1961).

Meynell, E., Datta, N.: (1) The relation of resistance transfer factors to the F-factor of *Escherichia coli* K12. Genet. Res. **7**, 134—140 (1966).

— — (2) The nature and incidence of conjugation factors in *Escherichia coli.* Genet. Res. **7**, 141—148 (1966).

— Meynell, G. G., Datta, N.: Phylogenetic relationship of drug-resistance factors and other transmissible bacterial plasmids. Bacteriol. Rev. **32**, 55—83 (1968).

MEYNELL, G. G.: Exclusion, superinfection immunity and abortive recombinants in I⁺ x I⁺ bacterial crosses. Genet. Res. **13**, 113—115 (1969).

— EDWARDS, S.: Failure of Col I to integrate in the bacterial chromosome. J. Gen. Microbiol. **53**, xii (1969).

— LAWN, A. M.: Sex pili and common pili in the conjugational transfer of colicin factor Ib by *Salmonella typhimurium*. Genet. Res. **9**, 359—367 (1967).

MITUSI, E., MIZUNO, D.: Stabilization of colicin E2 by bovine serum albumen. J. Bacteriol. **100**, 1136—1137 (1969).

MIYAMI, A., ICHIKAWA, S., AMANO, T.: Spheroplasts of colicine K producing *Escherichia coli* K235 prepared by leucozyme C and by lysozyme. Biken's J. **2**, 177—185 (1959).

— OZAKI, M., AMANO, T.: Double colicinogenicity of *Escherichia coli* K235. Biken's J. **4**, 1—11 (1961).

MONDOLFO, U., CEPPELLINI, R.: Antibiosi e fasi batteriche. Boll. ist. sieroterap. milan. **29**, 231—236 (1950).

MONK, M., CLOWES, R. C.: (1) Transfer of the colicin I factor in *Escherichia coli* K12 and its interaction with the F fertility factor. J. Gen. Microbiol. **36**, 365—384 (1964).

— — (2) The regulation of colicin synthesis and colicin factor transfer in *Escherichia coli*. J. Gen. Microbiol. **36**, 385—392 (1964).

MOORE, H. B., PICKETT, M. J.: (1) The *Pseudomonas-Achromobacter* group. Can. J. Microbiol. **6**, 35—42 (1960).

— — (2) Organisms resembling *Alkaligens faecalis*. Can. J. Microbiol. **6**, 43—52 (1960).

MOTOKI, Y., USHIO, K., YOSHINO, S.: Colicine type of *Shigella sonnei* isolated from many outbreak cases in Yamagachi prefecture and its neighbouring prefectures. J. Japan. Assoc. Infect. Diseases. **41**, 436—444 (1968).

MUKAI, F. H.: Interrelationship between colicin sensitivity and phage resistance. J. Gen. Microbiol. **23**, 539—551 (1960).

MULCZYK, M., KRUKOWSKA, A., SLOPEK, S.: Typing of *Shigella sonnei* by colicin production. Arch. Immunol. Therap. Exp. **15**, 622—626 (1967).

— SLOPEK, S., MARCINOWSKA, H.: Phage typing, colicin sensitivity and colicinogeny of *Shigella flexneri* 6. Arch. Immunol. Therap. Exp. **15**, 609—611 (1967).

MURAYAMA, Y., KOTANI, S. KATO K.: Solubilization of phage receptor substances from cell walls of *Staphylococcus aureus* (strain Copenhagen) by cell wall lytic enzymes. Biken's J. **11** 269—291 (1968).

MURRAY, R. G. E., STEED, P., ELSON, H. E.: The location of the mucopeptide in sections of the cell wall of *Escherichia coli* and other gram-negative bacteria. Can. J. Microbiol. **11**, 547—560 (1965).

NAGEL DE ZWAIG, R.: Association between colicinogenic and fertility factors. Genetics. **54**, 381—390 (1966).

— Mode of action of colicin A. J. Bacteriol. **99**, 913—914 (1966).

— ANTON, D. N.: Interactions between colicinogenic factors and fertility factors. Biochem. Biophys. Res. Communs. **17**, 358—362 (1964).

— — Genetic aspects of colicinogeny. Nat. Cancer. Inst. Monograph. **18**, 53—64 (1965).

— — PUIG, J.: The genetic control of colicinogenic factors E2, I and V. J. Gen. Microbiol. **29**, 473—484 (1962).

— LURIA, S. E.: Genetics and physiology of colicin tolerant mutants of *Escherichia coli*. J. Bacteriol. **94**, 1112—1123 (1962).

— — New class of conditional colicin tolerant mutants. J. Bacteriol. **99**, 78—84 (1969).

— PUIG, J.: The genetic behaviour of colicinogenic factor E1. J. Gen. Microbiol. **36**, 311—321 (1964).

NAGY, E.: Studies on the optimum conditions of megacin production in synthetic culture media. Acta Microbiol. Acad. Sci. Hung. **5**, 399—404 (1958).

— Remarks to megacin resistance. Zentr. Bakteriol. Parasitenk. Abt. I. Orig. **196**, 329—330 (1965).

— ALFOLDI, L., IVANOVICS, G.: Megacins. Acta Microbiol. Acad. Sci. Hung. **6**, 327—336 (1959).

NAITO, T., KONO, M., FUJISE, N., YAKUSHIJI, Y., AOKI, Y.: Colicine typing of *Shigella sonnei*. I. Principle, technique, selection of indicator strains and foundations for a typing scheme. Japan. J. Microbiol. **10**, 13—22 (1966).

NICOLLE, P., PRUNET, J.: La propriété colicinogène dans l'espèce *Salmonella typhi*. Ann. Inst. Pasteur **107** (Suppl. to No. 5) 174—189 (1964).

NOMURA, M.: Mode of action of colicines. Cold Spring Harbor Symposia Quant. Biol. **28**, 315—324 (1963).

— Mechanism of action of colicines. Proc. Natl. Acad. Sci. US **52**, 1514—1521 (1964).

— Colicins and related bacteriocins. Ann. Rev. Microbiol. **21**, 257—284 (1967).

— HALL, B. D., SPIEGELMAN, S.: Characterization of RNA synthesised in *Escherichia coli* after bacteriophage infection. J. Mol. Biol. **2**, 306—326 (1960).

— HOSODA, J.: (1) The autolytic phenomenon of *Bacillus subtilis* III. Some properties of autolysin. j. Agr. Chem. Soc. Japan **30**, 233—237 (1956).

— — (1) Nature of the primary action of the autolysin of *B. subtilis*. J. Bacteriol. **75**, 573—581 (1956).

— MAEDA, A.: Mechanism of action of colicines. Zentr. Bakteriol. Parasitenk. Abt. I Orig. **196**, 216—239 (1965).

— MATSUBARA, K., OKAMOTO, K., FUJIMURA, R.: Inhibition of host nucleic acid and protein synthesis by bacteriophage T4: Its relation to the physical and functional integrity of host chromosome. J. Mol. Biol. **5**, 535—549 (1962).

— NAKAMURA, M.: Reversibility of inhibition of nucleic acids and protein synthesis by colicin K. Biochem. Biophys. Res. Communs. **7**, 306—309 (1962).

— WITTEN, C.: Interaction of colicins with bacterial cells. III. Colicin-tolerant mutations in *Escherichia coli*. J. Bacteriol. **94**, 1093—1111 (1967).

— — MATEI, N., ECHOLS, H.: Inhibition of host nucleic acid synthesis by bacteriophage T4: effect of chloramphenicol at various multiplicities of infection. J. Mol. Biol. **17**, 273—278 (1966).

NOSE, K., MIZUNO, D.: Degradation of ribosomes in *Escherichia coli* cells treated with colicin Ez. J. Biochem. (Tokyo) **64**, 1—6 (1968).

— — OZEKI, H.: Degradation of ribosomal RNA from *E. coli* induced by colicine E2. Biochem. et Biophys. Acta. **119**, 636—638 (1966).

NOVICK, R. P.: Extrachromosomal inheritance in bacteria. Bacteriol. Rev. **33**, 210—235 (1969).

NUSKE, R., HOSEL, G., VENNER, H., ZINNER, H.: Über ein colicin aus *Escherichia coli* SG 710. Biochem. Z. **329**, 346—360 (1957).

OBINATA, M., MIZUNO, D.: Mechanism of colicin E2-induced DNA degradation in *Escherichia coli*. Biochem. et Biophys. Acta **199**, 330—339 (1970).

OHKI, M., OZEKI, H.: Isolation of conjugation-constitutive mutants of colicin factor Ib. Mol. Gen. Genet. **103**, 37—41 (1968).

OKAMOTO, K., MUDD, J. A., MANGAN, J., HUANG, W. M., SUBBAIAH, T. V., MARMUR, J.: Properties of the defective phage of *Bacillus subtilis*. J. Mol. Biol. **34**, 413—428 (1968).

ONISHI, Y.: Effects of pyocin 28 on sensitive bacteria. Fukuoka Acta Med. **60**, 577—590 (1969).

ONODERA, K., ROLFE, B., BERNSTEIN, A.: Demonstration of missing proteins in deletion mutants of *E. coli* K12. Biochem. Biophys. Res. Commun. **39**, 969—975 (1970).

OSMAN, M. A. M.: Pyocin typing of *Pseudomonas aeruginosa*. J. Clin. Path. **18**, 200—202 (1965).

OZAKI, M., AMANO, T.: Immunity to megacin A in protoplasts of megacinogenic *Bacillus megaterium*. Biken's J. **10**, 23—24 (1967).

— HIGASHI, Y., SAITO, H., AN, T., AMANO, T.: Identity of megacin A with phospholipase A. Biken's J. **9**, 201—213 (1966).

OZEKI, H.: Colicinogeny in *Salmonella*. Ph. D. Thesis, London University 1960.

— The behaviour of colicinogenic factors in *Salmonella typhimurium*. Zentr. Bakteriol. Parasitenk. Abt. I. Orig. **196**, 160—173 (1965).

— HOWARTH, S.: Colicine factors as fertility factors in bacteria: *Salmonella typhimurium* strain LT2. Nature **190**, 986—988 (1961).

— STOCKER, B. A. D.: Transduction by phage of colicinogeny. Heredity **12**, 525 (1958).

— — DE MARGERIE, H.: Production of colicine by single bacteria. Nature **184**, 337—339 (1959).

— — SMITH, S. M.: Transmission of colicinogeny between strains of *Salmonella typhimurium* grown together. J. Gen. Microbiol. **28**, 671—687 (1962).

PAPAVASSILIOU, J.: (1) Colicinogénie et sensibilité aux colicines des *Salmonella typhimurium* isolées au Congo Belge. Ann. soc. belg. méd. trop. **40**, 369—372 (1960).
— (2) Colicinogénie et sensibilité aux colicines des *Escherichia* d'origines humaine et animale. Arch. inst. Pasteur. Tunis **37**, 103—111 (1960).
— (3) Résistances croisées aux colicines I et B chez un mutant auxotrophe d'*Escherichia coli* K12. Arch. inst. Pasteur Hellénique. **6**, 133—138 (1960).
— (1) Biological characteristics of colicine X. Nature **190**, 110 (1961).
— (2) Lysogeny and colicinogeny in *Escherichia coli*. J. Gen. Microbiol. **25**, 409—413 (1961).
— (3) Actions antibiotiques réciproques chez *Pseudomonas aeruginosa*. Arch. inst. Pasteur Tunis **38**, 57—63 (1961).
— Preservation of colicinogeny and sensitivity to colicines in egg medium or by lysophilization. Pathol. Microbiol. **25**, 144—152 (1962).
— Sensitivity of *Salmonella* to colicines. Israel J. Med. Sci. **1**, 627—629 (1965).
— HUET, M.: Essais de colicinotypie de *Shigella flexneri*. Arch. inst. Pasteur Tunis **39**, 327—340 (1962).
— SAMARAKI-LYBEROPOULOU, V.: Production of colicins by *Salmonella*. Sensitivity of *Salmonella* to 5 colicins. Acta Microbiol. Hellenica **2**, 325—329 (1957).
— VANDEPITTE, J., GATTI, F., DE MOOR, J.: Colicinogénie et sensibilité aux colicines de 83 souches de *Shigella sonnei* isolée au Congo. Ann. inst. Pasteur **106**, 255—266 (1964).
PARDEE, A. D., JACOB, F., MONOD, J.: The genetic control and cytoplasmic expression of 'inducibility' in the synthesis of $\beta$-galactosidase by *E. coli*. J. Mol. Biol. **1**, 165—178 (1959)
PARKER, M. T., SIMMONS, L. E.: The inhibition of *Corynebacterium dipththeriae* and other gram positive organisms by *Staphylococcus aureus*. J. Gen. Microbiol. **21**, 457—476 (1959).
PATERSON, A. C.: Bacteriocinogeny and lysogeny in the genus *Pseudomonas*. J. Gen. Microbiol. **39**, 295—303 (1965).
PFAFF, P., WHITNEY, E. N.: Map position of the mutation for colicin E resistance in *Escherichia coli*. Bacteriol. Proc. **1971**, 50.
PFAHL, M.: Escape synthesis of $\beta$-galactosidase in strains of *E. coli* lysogenic for $\lambda$Cl-857, h80t-68, dlac. Mol. Gen. Genet. **105**, 122—124 (1969).
PRINSLOO, H. E.: Bacteriocins and phages produced by *Serratia marcescens*. J. Gen. Microbiol. **45**, 205—212 (1966).
— MARE, I. J., COETZEE, J. N.: Agar electrophoresis of antibiotics produced by *Serratia marcescens*. Nature **206**, 1055 (1965).
PRITCHARD, R. H., LARK, K. G.: Induction of replication by thymine starvation at the chromosome origin in *Escherichia coli*. J. Mol. Biol. **9**, 288—307 (1964).
PUCK, T. T., LEE, H. H.: Mechanism of cell wall penetration by viruses II. Demonstration of cyclic permeability change accompanying virus infection of *Escherichia coli* B cells. J. Exp. Med. **101**, 151—175 (1955).
PUIG, J., NAGEL DE ZWAIG, R.: Étude génétique d'un facteur colicinogène B et son influence sur la fertilité des croisements chez *Escherichia coli* K12. Ann. inst. Pasteur **107** (Suppl. to No. 5), 115—131 (1964).
RAMPINI, C., SCHIRMANN, J., BARBU, E.: Selection par les colicines d'une classe particulière de mutants thermosensibles d'*Escherichia coli*. Compt. rend. **264**, 1660—1663 (1967).
RAPIN, A. M. C., KALCKAR, H. M., ALBERICO, L.: The metabolic basis for masking of receptor-sites on *E. coli* K12 C21, a lipopolysaccharide core specific phage. Arch. Biochem. Biophys. **128**, 95—105 (1968).
REEVES, P. R.: Preparation of a substance having colicin F activity from *Escherichia coli* C.A. 42. Australian J. Exp. Biol. Med. Sci. **41**, 163—170 (1963).
— (1) The adsorption and kinetics of killing by colicin CA42-E2. Australian J. Exp. Biol. Med. Sci. **43**, 191—200 (1965).
— (2) The bacteriocins. Bacteriol. Rev. **29**, 24—45 (1965).
— Mutants resistant to colicin CA42-E2; cross resistance and genetic mapping of a special class of mutants. Australian J. Exp. Biol. Med. Sci. **44**, 301—316. Erratum **45**, 330 (1966).
— Mode of action of colicins of types E1, E2, E3 and K. J. Bacteriol **96**, 1700—1703 (1968).
VAN RENSBURG, A. J., HUGO, N.: Characterisation of DNA of colicinogenic factor E1 in a *Providence* strain. J. Gen. Microbiol. **58**, 421—422 (1969).

REVEL, H. R., LURIA, S. E.: Biosynthesis of $\beta$-galactosidase controlled by phage-carried genes II. The behaviour of phage-transduced z$^+$ genes toward regulatory mechanisms. Proc. Natl. Acad. Sci. US **47**, 1968—1974 (1961).

REYNOLDS, B. L.: The mode of action of colicin CA42-E2. Ph. D. Thesis, University of Adelaide 1966.

— REEVES, P. R.: Some observations on the mode of action of colicine F. Biochem. Biophys. Res. Commun. **11**, 140—145 (1963).

— — Kinetics and adsorption of colicin CA42-E2 and reversal of its bactericidal activity. J. Bacteriol. **100**, 301—309 (1969).

RILEY, M., PARDEE, A. B., JACOB, F., MONOD, J.: On the expression of a structural gene. J. Mol. Biol. **2**, 216—225 (1960).

RINGROSE, P.: Sedimentation analysis of DNA degradation products resulting from the action of colicin E2 on *Escherichia coli*. Biochim. et Biophys. Acta **213**, 320—334 (1970).

ROLFE, B., BERNSTEIN, A., ONODERA, K., BECKER, A.: Pleiotropic characteristics of the membrane mutant TolD of *E. coli*. Bacteriol. Proc. **1971**, 50.

— ONODERA, K.: Demonstration of missing membrane proteins in a colicin-tolerant mutant of *E. coli* K12. Biochem. Biophys. Res. Commun. **44**, 767—773 (1972).

ROSATO, R. R., CAMERON, J. A.: The bacteriophage receptor sites of *Staphylococcus aureus*. Biochim. et Biophys. Acta **83**, 113—119 (1964).

ROTH, T. F., HELINSKI, D. R.: Evidence for circular DNA forms of a bacterial plasmid. Proc. Natl. Acad. Sci. US **58**, 650—657 (1967).

RUDE, E., GOEBEL, W. F.: Colicine K. V. The somatic antigen of a non-colicinogenic variant of *E. coli* K235. J. Exp. Med. **116**, 73—100 (1962).

RYAN, F. J., FRIED, P., MUKAI, F.: A colicin produced by cells that are sensitive to it. Biochim. et Biophys. Acta **18**, 131 (1955).

SABET, F. S., SCHNAITMAN, C. A.: Localization and solubilization of colicin receptors. J. Bact. **105**, 422—430 (1971).

SALTON, M. R.: The bacterial cell wall. Amsterdam: Elsevier 1964.

SAMARAKI-LYBEROPOULOU, V., PAPAVASSILIOU, J.: Heterogeneity of *Shigella* in regard to sensitivity to colicines. Zentr. Bacteriol. Parasitenk. Abt. I. Orig. **196**, 351—363 (1965).

SANDOVAL, H. K., REILLY, C. H., TANDLER, B.: Colicin 15: possibly a defective bacteriophage. Nature **205**, 522—523 (1965).

SASARMAN, A., ANTOHI, M.: Présence des bactériocines chez le *Cl. perfringens*. Arch. roumaines pathol. exptl. microbiol. **22**, 377—380 (1963).

SAXE, L. S., LURIA, S. E.: The effect of colicin E2 on lambda DNA supercoils in super-infected *E. coli* K12 lysogens. Bacteriol. Proc. **1971**, 50.

SCHWARTZ, S. A., HELINSKI, D. R.: Purification and characterization of colicin E1. Bacteriol. Proc. **1968**, 53.

SENIOR, B. W., HOLLAND, I. B.: Effect of colicin E3 upon the 30S ribosomal subunit of *Escherichia coli*. Proc. Natl. Acad. Sci. US **68**, 959—963 (1971).

— KWASNIAK, J., HOLLAND, I. B.,: Colicin E3-directed changes in ribosome function and polyribosome metabolism in *Escherichia coli* K12. J. Mol. Biol. **53**, 205—220 (1970).

SHANNON, R., HEDGES, A. J.: Kinetics of lethal adsorption of colicin E2 by *Escherichia coli*. J. Bacteriol. **93**, 1353—1359 (1967).

SICARD, N., DEVORET, R.: Effets de la carence en thymine sur des souches d'*Escherichia coli* lysogènes K12T- et colicinogène 15T- Compt. rend. **255**, 1417—1419 (1962).

SICARDI, A. G.: Colicin resistance associated with resistance factors in *Escherichia coli*. Genet. Res. **8**, 219—228 (1966).

SILVER, S., OZEKI, H.: Transfer of deoxyribonucleic acid accompanying the transmission of colicinogenic properties by cell mating. Nature **195**, 873—874 (1962).

SINSHEIMER, R. L.: The replication of viral RNA. Symposium Soc. Gen. Microbiol. **18**, 101—124 (1968).

SLOPEK, S., MARESZ-BABCZYSZYN, J.: A working scheme for typing *Klebsiella* bacilli by means of pneumocins. Arch. Immunol. Therap. Exp. **15**, 525—529 (1967).

— MULCZYK, M., KRUKOWSKA, A.: New typing scheme of *Shigella sonnei* by colicin production. Arch. Immunol. Therap. Exp. **15**, 627—630 (1967).

SMARDA, J.: Incidence and manifestations of colicinogeny in strains of *Escherichia coli*. J. Hyg., Epidemiol., Microbiol., Immunol. (Prague) **4**, 151—165 (1960).
— Induction of the formation of a coli-bacteriophage and colicin by hydroperoxide. Experientia **18**, 271 (1962).
— Remarks on the action of colicin on *E. coli* cells and spheroplasts. Zentr. Bakteriol. Parasitenk. Abt. I. Orig. **196**, 240—248 (1965).
— (1) Ein von *Proteus mirabilis* produzierter Inhibitor des Colicins G. Z. allgem. Mikrobiol. **6**, 339—360 (1966).
— (2) The question of common receptor of phage T6 and colicine K. Proc. XIth Conf. Charles Univ. Med. Faculty. Prague 1966.
— KOUDELKA, J., KLEINWACHTER, V.. Induction of bacteriophage and colicin by means of acridine orange. Experientia **20**, 500—501 (1964).
— OBDRZALEK, V.: Colicine Q. Zentr. Bakteriol. Parasitenk. Abt. I. Orig. **200**, 493—497 (1966).
— SCHUHMANN, E.: Do certain colicines and phages share common receptors. Nature **213**, 614 (1966).
— TAUBENECK, U.: Situation of colicin receptors in surface layers of bacterial cells. J. Gen. Microbiol. **52**, 161—172 (1968).
— VRBA, M.: The microscopic picture of cells and penicillin induced spheroplasts of *Escherichia coli* exposed to the action of colicin. Folia Microbiol (Prague) **7**, 104—108 (1962).
SMIT, J. A., DE KLERK, H. C., COETZEE, J. N.: Properties of a *Proteus morganii* bacteriocin. J. Gen. Microbiol. **54**, 67—75 (1968).
SMITH, C.: Ph. D. Thesis, Bristol University 1966.
SMITH, D. A., BURROWS, T. W.: Phage and bacteriocin investigations with *Pasteurella pestis* and other bacteria. Nature **193**, 397—398 (1962).
SMITH, H. O., LEVINE, M.: The synthesis of phage and host DNA in the establishment of lysogeny. Virology **25**, 585—590 (1965).
SMITH, S. M., OZEKI, H., STOCKER, B. A. D.: Transfer of col E1 and col E2 during high-frequency transmission of col I in *Salmonella typhimurium*. J. Gen. Microbiol. **33**, 231—242 (1963).
SOMERS, J. M., BEVAN, E. A.: The inheritance of the killer character in yeast. Genet. Res. **13**, 71—83 (1969).
STAVSKII, V. N., KNYSH, I. N., FOMICHEV, Y. K.: Colicinogenicity and sensitivity to colicins of the causative agents of dysentery isolated from patients having acute and chronic dysentery. Zhur. Mikrobiol. Epidemiol. Immunobiol. **45** (6), 153—154 (1968).
STENT, G.: Molecular biology of bacterial viruses. W. H. Freeman 1963.
STEPANKOVSKAYA, L. D., BRUTMAN, E. I.: Typing colicins produced by *Shigella sonnei*. Zhur. Mikrobiol. Epidemiol. Immunobiol. **45** (11), 108—111 (1968).
STOCKER, B.: Heterogeneity of I colicines and I colicine factors. Heredity **21**, 166 (1966).
STOCKER, B. A. D., SMITH, S. M., OZEKI, H.: High infectivity of *Salmonella typhimurium* newly infected by the col I factor. J. Gen. Microbiol. **33**, 201—221 (1963).
STOUTHAMER, A. H., TIEZE, G. A.: Bacteriocin production by members of the genus *Klebsiella*. Antonie van Leeuwenhoek. J. Microbiol. Serol. **32**, 171—182 (1966).
STROBEL, M., NOMURA, M.: Restriction of the growth of a bacteriophage BF23 by colicin I (Col I-P9) factor. Virology **28**, 763—765 (1966).
SUBBAIAH, T. V., GOLDTHWAITE, C. D., MARMUR, J.: Nature of bacteriophages induced in *Bacillus subtilis*. In: Evolving genes and proteins (BRYSON and VOGEL, Eds.) Academic Press 1965.
SWORD, C. P., PICKETT, M. J.: The isolation and characterization of bacteriophages from *Listeria monocytogenes*. J. Gen. Microbiol. **25**, 241—248 (1961).
SZYBALSKY, W., KUBINSKI, H., SHELDRICK, P.: Pyrimidine clusters on the transcribing strand of DNA and their possible role in the initiation of RNA synthesis. Cold. Spring Harbor Symposia Quant. Biol. **31**, 123—127 (1966).
TAIZO, N., OZEKI, H.: Early abortive lysis by phage BF23 in *Escherichia coli* K12 carrying the colicin Ib factor. J. Virol. **2**, 1249—1254 (1968).

TAKEYA, K., MINAMISHIMA, Y., AMAKO, K., OHNISHI, Y.: Rod shaped pyocin 28. J. Gen. Virol. **4**, 145—149 (1969).
— SHIMODORI, S.: New method for the detection of lethal factor in Vibrios. J. Bacteriol. **19**, 339 (1969).
TAUBENECK, U.: Über inkomplette Bacteriophagen aus defekten lysogenen *Proteus mirabilis* Stämmen. Biol. Zentr. **86** (suppl), 45—54 (1967).
TAYLOR, A. L.: Bacteriophage-induced mutation in *Escherichia coli*. Proc. Natl. Acad. Sci. US **50**, 1043—1051 (1963).
TAYLOR, K.: Physical and chemical changes of Vi-polysaccharide due to Vi-phage II action. Acta Biochim. Polon. **13**, 97—106 (1966).
— TAYLOR, A.: Estimation of Vi-receptor activity. Acta Microbiol. Polon. **12**, 97—106 (1963).
TERZI, M.: Studies on the mechanism of bacteriophage T4 interference with host metabolism J. Mol. Biol. **28**, 37—44 (1967).
— LEVINTHAL, C.: Effects of λ-phage infection on bacterial synthesis. J. Mol. Biol. **26**, 525—535 (1967).
THRELFALL, E. J., HOLLAND, I. B.: Co-transduction with SerB of a pleiotropic mutation affecting colicin E2 refractivity, ultraviolet sensitivity, recombination proficiency and surface properties of *Escherichia coli* K12. J. Gen. Microbiol. **62**, 383—398 (1970).
TOKIWA, H., SAKAMOTO, S., KAJIKURI, M.: Colicine typing and colicinogeny of *Shigella sonnei*. Japan. J. Bacteriol. **22**, 141—145 (1967).
TSAO, S., GOEBEL, W. F.: Colicin K. VII The immunological properties of mitomycin induced colicin K. J. Exp. Med. **130**, 1313—1335 (1969).
TUBYLEWICZ, H.: Studies on the bacteriocinogeny of *Listeria monocytogenes* strains. Bull. acad. polon. sci., Sér. Sci. Biol. **11**, 519—521 (1963).
— Experimental studies on bacteriocinogeneity in *Clostridium perfringens* type A. I Isolation of bacteriocines and their antibacterial spectrum. Bull. acad. polon. sci., Sér. Sci. Biol. **14**, 31—36 (1966).
UETAKE, H., LURIA, S. E., BURROWS, J. W.: Conversion of somatic antigens in *Salmonella* by phage infection leading to lysis or lysogeny. Virology **5**, 68—91 (1958).
VASSILIADIS, P.: Essais de colicinogénie et colicinotypie de *Shigellae* isolées en Grèce. Ann. soc. belg. méd. trop. **44**, 341—346 (1964).
— PAPAVASSILJOU, J., GLAUDOT, A., SARTIAUX, P.: Production de colicines chez les *Salmonella* (étude sur 410 souches). Ann. inst. Pasteur **99**, 926—929 (1960).
— PATERAKI, E., POLITI, G.: Essais de colicinotypie de *Shigella* isolées en Grèce. Arch. inst. Pasteur Hellénique **10**, 37—43 (1964).
VIDAVER, A. K., BROCK, T. D.: Purification and properties of a bacteriophage receptor material from *Streptococcus faecium*. Biochim. et Biophys. Acta **121**, 298—314 (1966).
VIEU, J. F.: Applications de la bactériocinogénie et des bactériocines. Ann. inst. Pasteur **197** (Suppl. to No. 5), 93—114 (1964).
— CROISSANT, O., HAMON, Y.: Ultrastructure d'un agent antibactérien produit par *Yersinia enterocolitica*. Compt. rend. **264** D, 181—184 (1967).
VOLKIN, E., ASTRACHAN, L.: Phosphorus incorporation in *Escherichia coli* ribosomes after infection with bacteriophage T2. Virology **2**, 149—161 (1956).
VOSTI, K. L.: Production and sensitivity to colicins among serologically classified strains of *Escherichia coli*. J. Bacteriol. **96**, 1947—1952 (1968).
WAHBA, A. H.: The production and inactivation of pyocines. J. Hyg. **61**, 431—441 (1963).
— (1) Hospital infection with *Pseudomonas pyocyanea*: an investigation by a combined pyocine and serological typing method. Brit. Med. J. **5427**, 86—89 (1965).
— (2) Pyocine typing of *Pseudomonas pyocyanea* and its relation to serological typing. Zentr. Bakteriol. Parasitenk. Abt. I Orig. **196**, 389—394 (1965).
— (3) Vibriocine production in the Cholera and El Tor vibrios. Bull. World Health Organization **33**, 661—664 (1965).
WAITE, W. M., FRY, B. A.: Effect of infection with phage lambda on the synthesis of protein, RNA and DNA in *Escherichia coli*. J. Gen. Microbiol. **34**, 413—426 (1964).
WARREN, R. J., BOSE, S. K.: Bacteriophage-induced inhibition of host functions. J. Virol. **2**, 327—334 (1968).

WATANABE, T., FUKASAWA, T.: Episome-mediated transfer of drug resistance in Enterobacteriaceae. IV. Interactions between resistance transfer factor and F-factor in *Escherichia coli* K12. J. Bacteriol. **83**, 727—735 (1962).

WEIDEL, W., KOCH, G., LOHSS, F.: Über die Zellmembran von *Escherichia coli* B. II. Der Rezeptorkomplex für die Bakteriophagen T3, T4 und T7. Vergleichende chemisch-analytische Untersuchungen. Z. Naturforsch. **9** b, 398—406 (1954).

WELTZIEN, H. U., JESAITIS, M. A.: Chemical nature of the colicin K receptor. Bacteriol. Proc. **1969**, 52.

— — The nature of the colicin K receptor of *Escherichia coli* Cullen. J. Exp. Med. **133**, 534—553 (1971).

WHITNEY, E. N.: The TolC locus in *Escherichia coli* K12. Genetics **67**, 39—53 (1971).

WILKINSON, R. G., STOCKER, B. A. D.: Genetics and cultural properties of mutants of *Salmonella typhimurium* lacking glucosyl or galactosyl lipopolysaccharide transferases. Nature **217**, 955—957 (1968).

WILLETTS, N. S., CLARK, A. J.: Characteristics of some multiply recombination deficient strains of *Escherichia coli*. J. Bacteriol. **100**, 231—239 (1969).

WOODS, D. R., BEVAN, E. A.: Studies on the nature of the killer factor produced by *Saccharomyces cerevisiae*. J. Gen. Microbiol. **51**, 115—126 (1968).

YOUNG, F. G.: Requirement for glucosylated teichoic acid for adsorption of phage in *Bacillus subtilis* 168. Proc. Natl. Acad. Sci. US **58**, 2377—2384 (1967).

# Subject Index

# Molecular Biology, Biochemistry and Biophysics